# Nano-Optics and Nanophotonics

The Springer Series in Nano-Optics and Nanophotonics provides an expanding selection of research monographs in the area of nano-optics and nanophotonics, science- and technology-based on optical interactions of matter in the nanoscale and related topics of contemporary interest. With this broad coverage of topics, the series is of use to all research scientists, engineers and graduate students who need up-to-date reference books. The editors encourage prospective authors to correspond with them in advance of submitting a manuscript. Submission of manuscripts should be made to the editor-in-chief, one of the editors or to Springer.

More information about this series at http://www.springer.com/series/8765

Takashi Yatsui
Editor

# Progress in Nanophotonics 5

Springer

*Editor*
Takashi Yatsui
School of Engineering
The University of Tokyo
Tokyo, Japan

ISSN 2192-1970 ISSN 2192-1989 (electronic)
Nano-Optics and Nanophotonics
ISBN 978-3-030-07476-0 ISBN 978-3-319-98267-0 (eBook)
https://doi.org/10.1007/978-3-319-98267-0


Softcover re-print of the Hardcover 1st edition 2018


This Springer imprint is published by the registered company Springer Nature Switzerland AG
The registered company address is: Gewerbestrasse 11, 6330 Cham, Switzerland

# Preface to *Progress in Nanophotonics*

As the first example, recent advances in photonic systems demand drastic increases in the degree of integration of photonic devices for large-capacity, ultrahigh-speed signal transmission and information processing. Device size has to be scaled down to nanometric dimensions to meet this requirement, which will become even more strict in the future. As the second example, photonic fabrication systems demand drastic decreases in the size of the fabricated patterns for assembling ultralarge-scale integrated circuits. These requirements cannot be met even if the sizes of the materials are decreased by advanced methods based on nanotechnology. It is essential to decrease the size of the electromagnetic field used as a carrier for signal transmission, processing, and fabrication. Such a decrease in the size of the electromagnetic field beyond the diffraction limit of the propagating field can be realized in optical near-fields. Nanophotonics, a novel optical technology that utilizes the optical near-field, was proposed by M. Ohtsu (the editor of this monograph series) in 1993 in order to meet these requirements. However, it should be noted that the true nature of nanophotonics involves not only its ability to meet the above requirements. It is also its ability to realize qualitative innovations in photonic devices, fabrication techniques, energy conversion, and information processing systems by utilizing novel functions and phenomena made possible by optical near-field interactions, which are otherwise impossible as long as the conventional propagating light is used. Based on interdisciplinary studies on condensed matter physics, optical science, and quantum field theory, nanomaterials and optical energy transfer in the nanometric regime have been extensively studied in the last two decades. Through these studies, novel theories on optical near-fields have been developed, and a variety of novel phenomena have been found. The results of this basic research have been applied to develop nanometer-sized photonic devices, nanometer-resolution fabrication, highly efficient energy conversion, and novel information processing, resulting in qualitative innovations. Further advancement in these areas is expected to establish novel optical sciences in the nanometric space, which can be applied to further progress in nanophotonics in order to support the sustainable development of peoples' lives all over the world. This unique monograph series entitled "Progress in Nanophotonics" in the Springer Series in

Nano-Optics and Nanophotonics is being introduced to review the results of advanced studies in the field of nanophotonics and covers the most recent topics of theoretical and experimental interest in relevant fields, such as classical and quantum optical sciences, nanometer-sized condensed matter physics, devices, fabrication techniques, energy conversion, information processing, architectures, and algorithms. Each chapter is written by leading scientists in the relevant field. Thus, this monograph series will provide high-quality scientific and technical information to scientists, engineers, and students who are and will be engaged in nanophotonics research. As compared to the previous monograph series entitled "Progress in Nano-Electro-Optics" (edited by M. Ohtsu, published in the Springer Series in Optical Science), this monograph series deals not only with optical science on the nanometer scale but also with its applications in technology. I am grateful to Dr. C. Ascheron of Springer-Verlag for his guidance and suggestions throughout the preparation of this monograph series.

Tokyo, Japan Motoichi Ohtsu

# Preface to Volume V

This book focuses on various and mutually related topics in nanophotonics written by the world's leading experts. Following the elaboration of the idea of nanophotonics, much theoretical and experimental work has been carried out, and several novel photonic devices, high-resolution fabrication, highly efficient energy conversion, and novel information processing were developed in these years. Novel theoretical models describing the nanometric light–matter interaction were also developed. Chapter 1 describes the historical review of dressed photons (DPs) carried out in the last three decades. After a review of the present status of experimental studies, required theories of DP are reviewed. Chapter 2 is devoted to describing a unique feature by spacelike momentum which describes field interactions into consideration, and it provides an extended picture of conventional virtual photon dynamics on spacetime. Chapter 3 describes a new framework of DP based on quantum probability, revealing that a probability measure plays crucial roles in nanophotonics. Chapter 4 is devoted to review many interesting aspects of symmetry breaking. Chapter 5 reviews quantum measurement theory to develop for DP. Chapter 6 describes the response theory for DP to overcome the difficulties due to the many-electron problem. This volume is a valuable resource for readers and future specialists in nanophotonics. The work reviewed in this book was partially supported by the JSPS Core-to-Core Program (A. Advanced Research Networks).

Tokyo, Japan — Takashi Yatsui

# Contents

# Contributors

**Itsuki Banno** Graduate Faculty of Interdisciplinary Research Faculty of Engineering, University of Yamanashi, Kofu, Yamanashi, Japan

**Motoichi Ohtsu** Research Origin for Dressed Photon, Yokohama, Kanagawa, Japan

**Izumi Ojima** Research Origin for Dressed Photon, Yokohama, Kanagawa, Japan

**Kazuya Okamura** Graduate School of Infomatics, Nagoya University, Nagoya, Aichi, Japan

**Hayato Saigo** Nagahama Institute of Bio-Science and Technology, Nagahama, Japan

**Hirofumi Sakuma** Research Origin for Dressed Photon, Yokohama, Kanagawa, Japan

# Chapter 1
# Historical Review of Dressed Photons: Experimental Progress and Required Theories

**Motoichi Ohtsu**

**Abstract** First, this article reviews the experimental and theoretical studies of dressed photons (DPs), carried out in the last three decades. It is pointed out that future theoretical studies can be developed by following three steps. Second, it is proposed how to proceed in the first and second steps by describing a unique DP energy transfer process. As an application of this transfer process, novel functional devices are demonstrated. The DP energy transfer in these devices is shown to exhibit an autonomous spatial evolution feature. A temporal evolution feature is also revealed. Another application is a highly efficient optical energy conversion film that gives a drastically increased electrical power generation efficiency of a silicon solar battery. Lastly, in order to identify the requirements for future theoretical studies, this article reviews how to proceed in the third step by employing a fiber probe. Among these requirements, an essential requirement is to describe the autonomy mentioned above and also the hierarchy observed in DP measurements. To meet the requirements, novel theoretical approaches are reviewed for developing a new field known as off-shell science. These approaches are based on Clebsch dual field theory, the quadrality scheme in category theory, and a novel measurement theory.

## 1.1 Introduction

The dressed photon (DP), a novel form of photon created in a nanometer-sized space, has been referred to as an optical near field, and the science for dealing with this type of photon has been called near field optics. The history of near field optics is long and can be classified into older and modern times, based on the great differences in the concepts, principles, and methods of studying the DP. The older time started

M. Ohtsu (✉)
Research Origin for Dressed Photon, c/o Nichia Corp., 3-13-19 Moriya-cho, Kanagawa-ku, Yokohama, Kanagawa 221-0022, Japan
e-mail: ohtsu@rodrep.or.jp



T. Yatsui (ed.), *Progress in Nanophotonics 5*, Nano-Optics and Nanophotonics,
https://doi.org/10.1007/978-3-319-98267-0_1

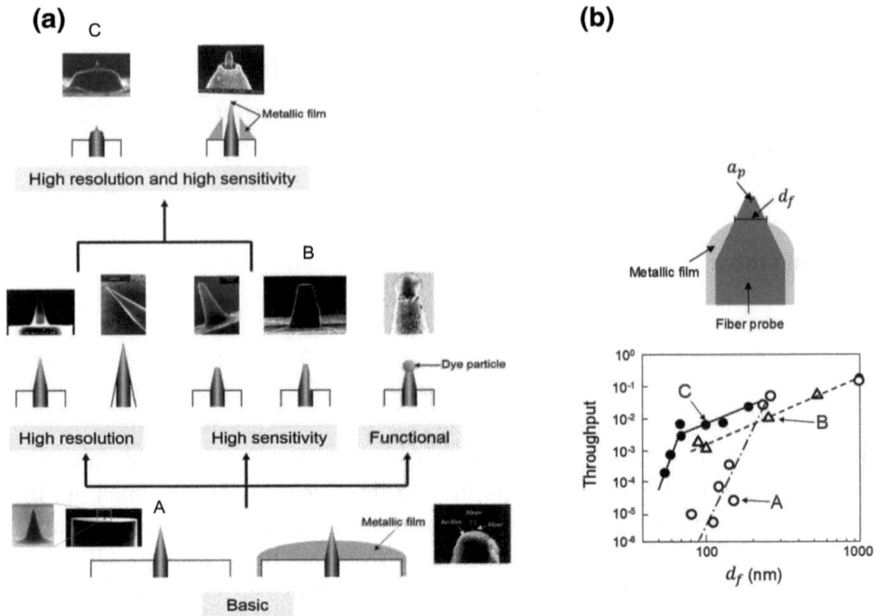


**Fig. 1.1** Developed fiber probes. **a** Cross-sectional profiles and scanning electron microscopic images. **b** The efficiency of measuring the DP (refer to Sect. 1.8.2). $d_f$ is the foot diameter of the taper protruding from an opaque metallic film. Open circles, open triangles, and closed circles represent the experimental results measured for the fiber probes A, B, and C in **a**, respectively

with a simple proposal to use light falling on a sub-wavelength sized aperture for high-resolution microscopy [1]. After a long time during which this proposal was ignored, theoretical analyses were carried out on the diffraction and radiation of electromagnetic waves through a small aperture [2, 3]. Afterward, these analyses were experimentally demonstrated by using microwaves [4].

Demonstrations using light were finally carried out using a novel methodology, named near-field optical microscopy, by several scientists in several countries, including Japan (M.O.), almost simultaneously [5]. Among them, this author (M.O.) developed high-quality optical fiber probes for generating and detecting optical near fields with high resolution and high sensitivity (Fig. 1.1) [6]. These fiber probes have been used to assemble novel spectrometers, and microscopic and spectroscopic images of specimens, e.g., a single strand of DNA, were successfully acquired with a high resolution beyond the diffraction limit [7]. They have since become commercially available and have been exported to many countries around the world [8].

Based on the successful experimental demonstrations above, the Near-Field Optics Workshop has held in order to promote basic studies and applications of optical near fields [5]. In this workshop, a physical picture of the optical near field was drawn using a conventional optical method, i.e., by using the dispersion relation between the momentum and energy of the photon. After this workshop, the

International Near-Field Optics Conference was organized, and the most recent 14th conference was held in 2016 [9]. In order to promote near field optics in the Asia-Pacific area, the author organized the Asia-Pacific Workshop on Near-Field Optics in 1996 [10]. After this workshop, the Asia-Pacific Near-Field Optics Conference was also organized, and the most recent 11th conference was held in 2017 [11].

It should be pointed out that the older time mentioned above has already ended. The reason for this is that conventional optical microscopy is based on the methodology of nondestructive measurement of the specimen's conformation and/or structure, which is described by conventional optics. Unlike this conventional microscopy, near-field optical microscopy is based on destructive measurement because it acquires an image of the specimen through optical near field interactions between the probe tip and the specimen. Even though a high resolution beyond the diffraction limit of light could be realized by this microscopy, a fatal problem was that electronic energy levels in the specimen were disturbed as a result of these interactions, which resulted in the acquired image profile being different from and uncorrelated with the conventional optical microscope images. Noting this difference, the application of the optical near field to microscopy ended, and the modern time started. In the modern time, enormous efforts were devoted to studying the above-mentioned optical near field interactions, i.e., light-matter interactions that take place via the optical near field [12].

With the aim of identifying routes between historical DP studies and future studies, first, Sects. 1.2–1.4 review the experimental and theoretical studies carried out in the last three decades [13]. It is pointed out that such routes can be developed by following three steps (steps 1–3). Sections 1.5–1.7 are devoted to proposing how to proceed with steps 1 and 2 by describing a unique DP energy transfer process [14]. Sections 1.8 and 1.9 review how to proceed with the third step [15] and identify requirements for a novel theory, leading to the development of a new generic science, named off-shell science. Section 1.10 is a summary of this article.

## 1.2 History of the Modern Time

As was pointed out in the previous section, there were great differences between studies on the DP in the older and modern times in terms of their concepts, principles, and methods. That is, the studies promoted in the modern time are essentially incompatible with those in the older time. They are called off-shell and on-shell sciences, respectively, as will be reviewed in Sect. 1.4. In the modern time, extensive experimental studies were carried out by using two kinds of materials. The first kind was metals, in which plasmonic oscillation of the free electrons was utilized. Although the ease of fabricating nanometer-sized metallic particles or metallic thin films was high, these materials had two intrinsic features; a short conversion time from the optical energy to the plasmonic oscillation energy of the electrons, and a short phase-relaxation time of the electrons. Due to these short time constants, the quantum nature of the incident light was not maintained in the metal. Therefore, quantum field theory was not required for describing the light-matter interactions in the metal,; instead, the conventional optical theory was sufficient, and conventional parameters, including the optical refractive index, the wave-number, and the mode, were used. This means that the dispersion relation was valid for describing the interactions, as has been used since the Near-Field Optics Workshop a quarter of a century ago. This situation means that studies using the first kind of material were left behind by the modern time.

Productive experimental studies in the modern time were promoted by using the second kind of materials, including semiconductors, organic materials, and gaseous molecules, in which their discrete electronic energy levels were utilized. Although fabrication of nanometer-sized particles or thin films was not straightforward in the beginning, technological advances enabled fabrication of suitable devices, which promoted further studies. As a result, a variety of applications were developed to establish a generic technology, as illustrated in Fig. 1.2 [16]. These applications include: lithography [17] and information storage [18] at densities beyond the diffraction limit, polishing of material surfaces to atomic-level flatness [19], photon breeding devices [20], functional devices (named DP devices) [21], optical router systems [22], optical pulse shape measurement systems [23], optical security systems [24], and energy conversion systems [25, 26].

To promote experimental studies in the modern time, three theoretical problems had to be solved:

[1] Problem 1: Quantum field theory for photons and electron-hole pairs was required to describe the light-matter interactions taking place in a nanometric space. However,

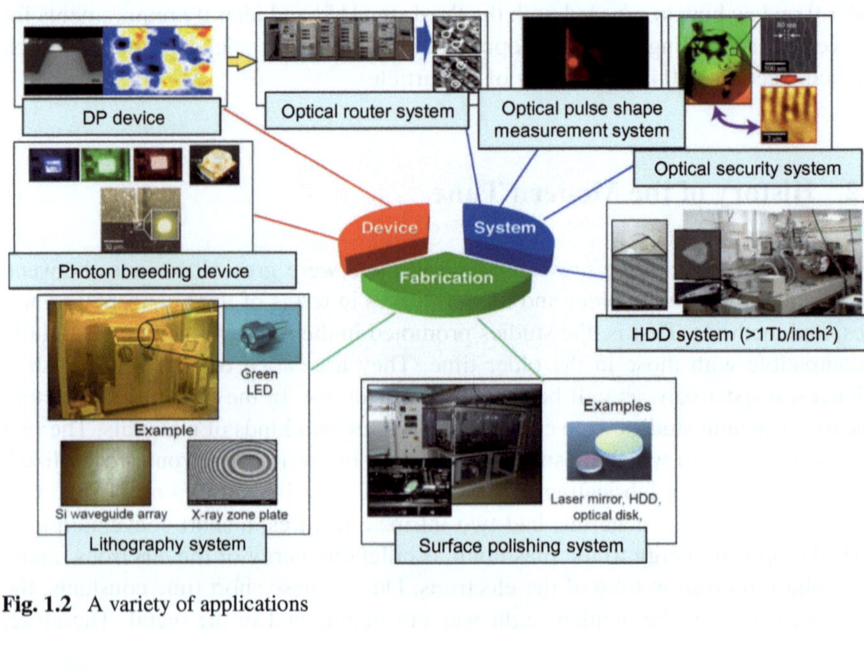


**Fig. 1.2** A variety of applications

the problem was that a virtual cavity, for deriving the Hamiltonian operator, could not be defined because the sub-wavelength-sized optical near field existed in a nanometric space. In other words, the electromagnetic mode could not be defined. It should be noted that the optical near field was regarded neither as a collection of corpuscles nor as a free photon, proposed by Newton and Einstein, respectively.

To solve this problem, a novel physical picture of the optical near field was drawn by representing the Hamiltonian operator $\hat{H}$ under illumination with propagating light having photon energy $\hbar\omega_o$:

$$\hat{H} = \sum_{k\lambda} \hbar\omega_k \hat{a}^\dagger_{k\lambda}\hat{a}_{k\lambda} + \sum_{\alpha>F,\beta<F} (E_\alpha - E_\beta)\hat{b}^\dagger_{\alpha\beta}\hat{b}_{\alpha\beta} + \hat{H}_{\text{int}}. \tag{1.1}$$

The first term represents the photon energy created in the nanometric space, which is given by the sum of an infinite number of photon modes with angular frequency $\omega_{\boldsymbol{k}}$, polarization state $\lambda$, and energy $\hbar\omega_{\boldsymbol{k}}$. Here, the subscript $\boldsymbol{k}$ represents the wave-vector, and $\hat{a}_{\boldsymbol{k}\lambda}$ and $\hat{a}^\dagger_{\boldsymbol{k}\lambda}$ are annihilation and creation operators, respectively. They satisfy the commutation relation

$$\left[\hat{a}_{\boldsymbol{k}\lambda}, \hat{a}^\dagger_{\boldsymbol{k}'\lambda'}\right] = \delta_{\boldsymbol{k}\boldsymbol{k}'}\delta_{\lambda\lambda'}, \tag{1.2}$$

where $\delta_{\boldsymbol{k}\boldsymbol{k}'}$ and $\delta_{\lambda\lambda'}$ are Kronecker deltas. The second term represents the energy of the electron-hole pair, which is also given by the sum of the energies of the electron-hole pairs of the infinite number of energy levels, identified by the subscripts $\alpha$ and $\beta$. The energy difference $E_\alpha - E_\beta$ represents the bandgap energy in the case of a semiconductor, and $F$ represents the Fermi energy level. The operators $\hat{b}_{\alpha\beta}$ and $\hat{b}^\dagger_{\alpha\beta}$ respectively represent the simultaneous annihilation and creation of the electron and hole, i.e., the annihilation and creation operators of the electron-hole pair. They satisfy the commutation relation

$$\left[\hat{b}_{\alpha\beta}, \hat{b}^\dagger_{\alpha'\beta'}\right] = \delta_{\alpha\alpha'}\delta_{\beta\beta'}. \tag{1.3}$$

The third term represents the energy of the interaction between the photon and the electron-hole pair, which is given by

$$\hat{H}_{\text{int}} = -\int \hat{\psi}^\dagger(\boldsymbol{r})\boldsymbol{p}(\boldsymbol{r})\hat{\psi}(\boldsymbol{r}) \cdot \hat{\boldsymbol{D}}^\perp(\boldsymbol{r})dv, \tag{1.4}$$

where $\boldsymbol{p}(\boldsymbol{r})$ is an electric dipole moment, $\hat{\psi}(\boldsymbol{r})$ and $\hat{\psi}^\dagger(\boldsymbol{r})$ are respectively annihilation and creation operators for the field of the electron-hole pair, and $\hat{\boldsymbol{D}}^\perp(\boldsymbol{r})$ is the transverse component of the electric displacement operator of the incident propagating light, which is perpendicular to the wave-vector $\boldsymbol{k}$. This operator $\hat{\boldsymbol{D}}^\perp(\boldsymbol{r})$ is expressed as

$$\hat{\boldsymbol{D}}^{\perp}(\boldsymbol{r}) = i \sum_{\boldsymbol{k}} \sum_{\lambda=1}^{2} N_{\boldsymbol{k}} \boldsymbol{e}_{\boldsymbol{k}\lambda}(\boldsymbol{k}) \left\{ \hat{a}_{\boldsymbol{k}\lambda}(\boldsymbol{k}) e^{i\boldsymbol{k}\cdot\boldsymbol{r}} - \hat{a}^{\dagger}_{\boldsymbol{k}\lambda}(\boldsymbol{k}) e^{-i\boldsymbol{k}\cdot\boldsymbol{r}} \right\}, \tag{1.5}$$

where plane waves are used for the mode functions. Here, $N_{\boldsymbol{k}}$ and $\boldsymbol{e}_{k\lambda}(\boldsymbol{k})$ are a proportionality constant and the unit vector along the direction of polarization, respectively.

By diagonalizing the Hamiltonian operator of (1.1), annihilation and creation operators ($\hat{a}$ and $\hat{a}^{\dagger}$, respectively) of the novel quasi-particle were derived, which represented the quantum state of the coupled photon and electron-hole pairs as a result of their interaction in the nanometric space:

$$\hat{a} = \sum_{\boldsymbol{k}\lambda} \left\{ \hat{a}_{\boldsymbol{k}\lambda} - i N_{\boldsymbol{k}} \sum_{\alpha > F, \beta < F} \left( \rho^{*}_{\alpha\beta\lambda}(\boldsymbol{k}) \hat{b}_{\alpha\beta} + \rho^{*}_{\beta\alpha\lambda}(\boldsymbol{k}) \hat{b}^{\dagger}_{\alpha\beta} \right) \right\}, \tag{1.6}$$

and

$$\hat{a}^{\dagger} = \sum_{\boldsymbol{k}\lambda} \left\{ \hat{a}^{\dagger}_{\boldsymbol{k}\lambda} + i N_{\boldsymbol{k}} \sum_{\alpha > F, \beta < F} \left( \rho_{\alpha\beta\lambda}(\boldsymbol{k}) \hat{b}^{\dagger}_{\alpha\beta} + \rho_{\beta\alpha\lambda}(\boldsymbol{k}) \hat{b}_{\alpha\beta} \right) \right\}, \tag{1.7}$$

where $\rho_{\alpha\beta\lambda}(\boldsymbol{k})$ is the spatial Fourier transform of the electric dipole moment. These operators are represented by the sum of the operators for the photons of the infinite number of modes and for the electron-hole pairs of the infinite number of energy levels. Because of this summation, this quasi-particle was named the dressed photon (DP), i.e., a photon dressed by the material excitation energy (Fig. 1.3) [27].

[2] Problem 2: Since the DP is non-propagating and is localized on a nanometer-sized material (nanoparticle: NP), another NP is required for its detection; in other words, another NP had to be placed in close proximity to convert the DP to propagating light via multiple scattering of the DP. Here, the problem was how to describe this scattering, because the two NPs, i.e., the source and detector for the DP, are not independent of each other but are coupled via the DP. However, this problem was solved by using the solution to Problem 1. That is, scattering was described by using the annihilation and creation operators of DPs on the two NPs.

[3] Problem 3: Since the actual NPs and DP are always surrounded by a macroscopic system composed of macroscopic materials and macroscopic optical fields, the prob-

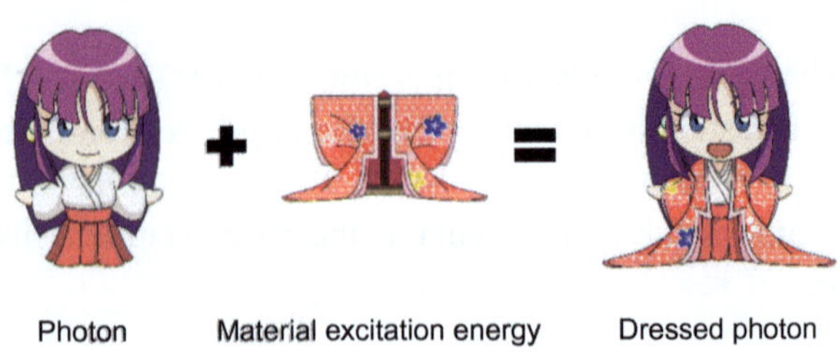


**Fig. 1.3** Illustrative explanation of the dressed photon

lem was how to take into account the contribution from the macroscopic energy in order to derive the magnitude of the DP interaction energy between NPs. To solve this problem, the contributions from the macroscopic system were renormalized by the projection operator method. As a result, the magnitude of the DP interaction energy was derived and represented by a Yukawa function:

$$Y(r) = \frac{\exp(-r/a)}{r}, \tag{1.8}$$

where $r$ is the distance from the center of the NP, and $a$ represents the extent of localization, which is equivalent to the size of the NP [28].

In the solution to Problem 3 above, it should be noted that $a$ is independent of the wavelength $\lambda$ of the incident propagating light. Furthermore, the DP is strongly localized in the sub-wavelength-sized space because $a \ll \lambda$. Due to this localization, the following two novel phenomena were found: The first one is with respect to the electronic transition. For explaining this transition, an atom with a simple two-energy level electron is considered by assuming that the state functions of these levels have the same parity. In this case, the electric dipole transition is forbidden under irradiation with propagating light incident on the atom. This is because the state functions have the same parity and, more essentially, the long-wavelength approximation is valid. However, this electric dipole-forbidden transition turns out be allowed in the case where the DP is involved. This is because the long-wavelength approximation is not valid any more since $a \ll \lambda$. The second phenomenon was named size-dependent resonance; in this phenomenon, the magnitude of the DP energy transferred between the two NP takes a maximum when the sizes of these NP are equal [29]. This corresponds to the momentum conservation law for the DP. It is different from diffraction, which is a typical phenomenon in classical optics, where the cross-sectional size of the light beam on a screen after being transmitted through an aperture is inversely proportional to the aperture size. By utilizing these two phenomena, a variety of applications were developed, some of which are shown in Fig. 1.2. To realize more developments, further theoretical studies were carried out.

One successful result of these studies was the further possibility of dressing, i.e., the possibility of coupling DPs and phonons [30]. In order to describe this coupling, for simplicity a nanomaterial tip is assumed, in which a one-dimensional crystal lattice is provided (Fig. 1.4a). When the tip is illuminated with light, a DP can be created on a lattice site and hops to an adjacent site. During the hopping, the DP can excite lattice vibrations to create phonons, and then the DP couples with these phonons.

For a theoretical formulation of this coupling, the Hamiltonian operator $\hat{H}$ is given by on:

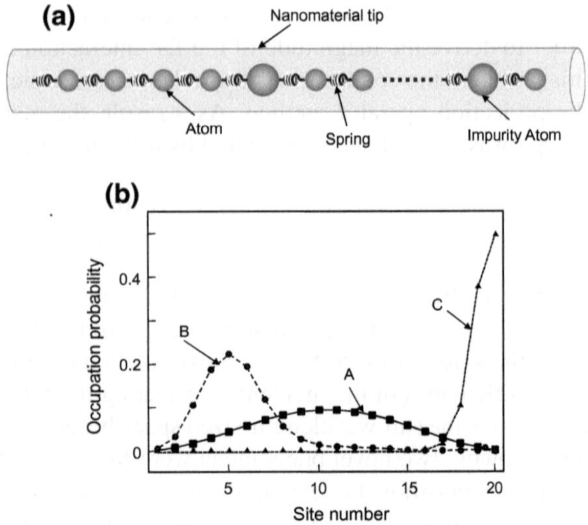


**Fig. 1.4** Occupation probability of the dressed photon. **a** One-dimensional model for calculation. The number of atoms is 20. The impurity atoms are at sites 4, 6, 13, and 19. **b** Calculated results. The mass of the impurity atoms is 0.2-times that of the other atoms. $\hbar\omega = 1.81\,\text{eV}$ and $\hbar J = 0.5\,\text{eV}$. Curves A, B, and C represent the results for $\chi_{ip} = 0, 40.0$, and 54.0 fs$^{-1}$nm$^{-1}$, respectively

$$\hat{H} = \sum_{i=1}^{N} \hbar\omega \tilde{a}_i^\dagger \tilde{a}_i + \sum_{p=1}^{N} \hbar\Omega_p \hat{c}_p^\dagger \hat{c}_p + \sum_{i=1}^{N}\sum_{p=1}^{N} \hbar\chi_{ip} \tilde{a}_i^\dagger \tilde{a}_i (\hat{c}_p^\dagger + \hat{c}_p) + \sum_{i=1}^{N-1} \hbar J (\tilde{a}_i^\dagger \tilde{a}_{i+1} + \tilde{a}_{i+1}^\dagger \tilde{a}_i). \quad (1.9)$$

In the first term, $\tilde{a}_i$ and $\tilde{a}_i^\dagger$ respectively denote the annihilation and creation operators of a DP with energy $\hbar\omega$ at site $i$ in the lattice. In the second term, $\hat{c}_p$ and $\hat{c}_p^\dagger$ are respectively the annihilation and creation operators of the phonon of mode $p$, which satisfies the boson commutation relation:

$$[\hat{c}_p, \hat{c}_q^\dagger] \equiv \hat{c}_p \hat{c}_q^\dagger - \hat{c}_q^\dagger \hat{c}_p. = \delta_{pq}. \quad (1.10)$$

The phonon energy is represented by $\hbar\Omega_p$. The third and fourth terms stand for the DP-phonon interaction with the interaction energy $\hbar\chi_{ip}$ and DP hopping with hopping energy $\hbar J$, respectively.

By diagonalizing $\hat{H}$ of (1.9), annihilation and creation operators of the novel quasi-particle are derived and expressed as

$$\hat{\alpha}_i = \tilde{a}_i \exp\left\{ \sum_{p=1}^{N} \frac{\chi_{ip}}{\Omega_p} (\hat{c}_p^\dagger - \hat{c}_p) \right\}, \quad (1.11)$$

$$\hat{\alpha}_i^\dagger = \tilde{a}_i^\dagger \exp\left\{-\sum_{p=1}^{N} \frac{\chi_{ip}}{\Omega_p}(\hat{c}_p^\dagger - \hat{c}_p)\right\}, \tag{1.12}$$

which is the product of the DP operator of (1.6) and (1.7) and the displacement operator function for the phonon. It should be noted that this function creates a multi-mode phonon with a coherent state. In other words, the DP excites a multi-mode coherent phonon, and they coupled to form a novel quasi-particle named a dressed-photon-phonon (DPP). Further theoretical studies found that the created DPP was localized on an impurity atom in a lattice site or on the end of the nanomaterial tip when the DP-phonon interaction energy was sufficiently high (curve B or C in Fig. 1.4b, respectively).

As a result of the theoretical studies on the DPP above, a novel light-matter interaction, named a DPP-assisted light-matter interaction, was found [31]: For simplicity, two electronic energy levels ($|E_u\rangle$ and $|E_l\rangle$) are assumed in the two NPs, and an electric dipole transition is allowed between these levels (Fig. 1.5). It is also assumed that the photon energy $h\nu_{in}$ of the incident light is lower than the energy difference $E_d = E_u - E_l$ between the energies of these two levels. Here, by irradiating a NP 1 with incident light, the DPP is created. When another NP 2 is installed in close

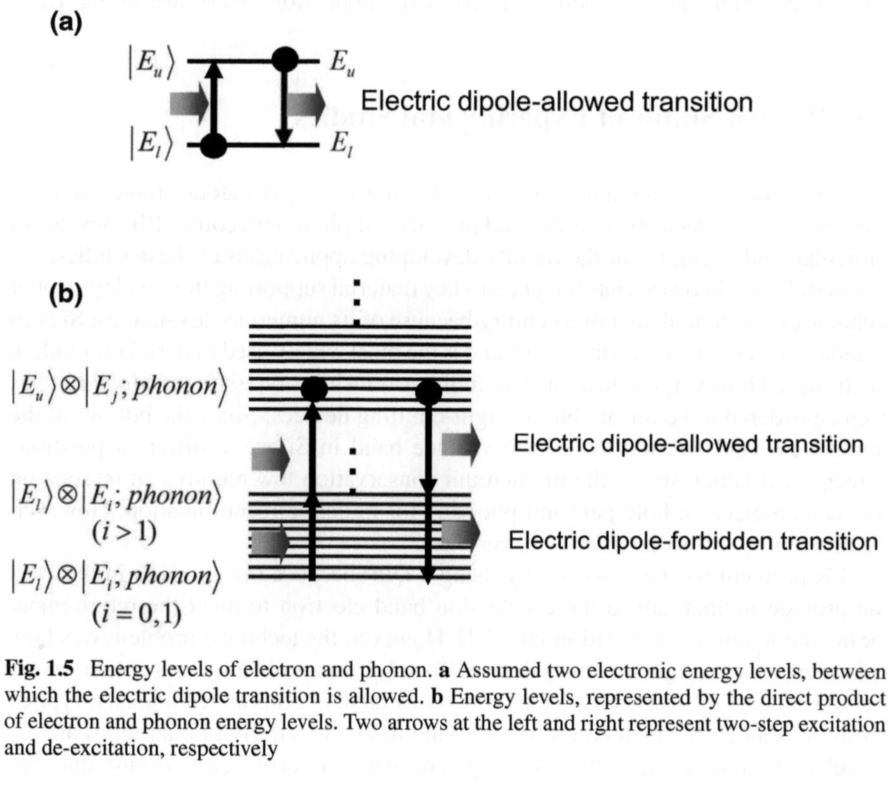


**Fig. 1.5** Energy levels of electron and phonon. **a** Assumed two electronic energy levels, between which the electric dipole transition is allowed. **b** Energy levels, represented by the direct product of electron and phonon energy levels. Two arrows at the left and right represent two-step excitation and de-excitation, respectively

proximity to NP 1, the DPP energy is transferred to NP 2. Here, to describe the novel light-matter interaction between the NPs 1 and 2, not only the two electronic energy levels but also infinite numbers of phonon energy levels have to be considered. This is because the DPP is a quasi-particle in which the DP is accompanied by multi-mode coherent phonons. That is, for describing the light-matter interaction induced in NP 2 by the DPP energy transfer, the electronic energy levels in NP 2 are accompanied by an infinite number of phonon energy levels. Thus, the energy levels in NP 2 are represented by $|E_l\rangle \otimes |E_i;\, phonon\rangle$ and $|E_u\rangle \otimes |E_i;\, phonon\rangle$ $(i = 1, 2, 3, \ldots)$, where $\otimes$ is the direct product. Thus, even though $h\nu_{in} < E_d$, NP 2 can be excited to a high phonon energy level $|E_l\rangle \otimes |E_i;\, phonon\rangle$ $(i > 1)$ by the first arriving incident photon. Here, it should be noted that this is an electric dipole-forbidden transition because it is a transition in the electronic energy state $|E_l\rangle$. In the case where $h\nu_{in} > E_d/2$, NP 2 is subsequently excited from $|E_l\rangle \otimes |E_i;\, phonon\rangle$ $(i > 1)$ to one of the phonon energy levels in the excited electronic energy level $|E_u\rangle \otimes |E_j;\, phonon\rangle$ by the second arriving incident photon. This transition is an electric dipole-allowed transition because it is the transition from $|E_l\rangle$ to $|E_u\rangle$. As a result of this two-step transition, a free electron is created in NP 2. The opposite transition is possible; i.e., the electron in level $|E_u\rangle \otimes |E_j;\, phonon\rangle$ is de-excited to $|E_l\rangle \otimes |E_i;\, phonon\rangle$ $(i = 0, 1)$ by the subsequent electric dipole-allowed and -forbidden transitions, and two photons are emitted. This novel DPP-assisted light-matter interaction has contributed considerably to the development of a variety of applications, as shown in Fig. 1.2.

## 1.3 Present Status of Experimental Studies

Recently, there have been a large number of extensive experimental studies on DPs. This section reviews the principles and practices of photon breeding (PB) devices in particular, and examples of the rapidly developing applications of these studies.

Crystalline silicon (Si) has long been a key material supporting the development of technology for more than half a century because of its numerous advantages: Si is an abundant material in the earth's crust, and is the most widely used material for modern electronics. However, because Si is an indirect-transition-type semiconductor, it has been considered to be unsuitable for light-emitting devices: Since the bottom of the conduction band and the top of the valence band in Si are at different positions in reciprocal lattice space, the momentum conservation low requires an interaction between an electron-hole pair and phonons for radiative recombination. However, the probability of this interaction is very low.

This problem has been solved by using a DPP because the phonons in the DPP can provide momentum to the conduction band electron to meet the requirement for the momentum conservation law [32]. However, the technical problem was how to fabricate such a light-emitting device. To solve this problem, a novel fabrication method named DPP-assisted annealing was invented. For this annealing, an n-type Si substrate is used, in which As atoms or Sb atoms are doped. By implanting B atoms, the substrate surface is transformed to a p-type material, forming a p-n homojunction.

After metallic films are coated to serve as electrodes, a forward current is injected. The principle of the DPP-assisted annealing is: By this current injection, Joule heat is generated to diffuse the B atoms. During this Joule-annealing, the substrate surface is irradiated with infrared light (for example, light with a wavelength of 1.3 μm). Because its photon energy $h\nu_{anneal}$ (=0.95 eV) is sufficiently lower than the bandgap energy $E_g$ (=1.12 eV) of Si, the light can penetrate into the Si substrate without suffering absorption. Then, the light reaches the p-n homojunction to create the DPP on the B atom. The created DPP localizes at this impurity atom, as explained by curve B in Fig. 1.4. Then, phonons in the created DPP can provide momenta to the electron nearby to satisfy the momentum conservation law, resulting in emission of a photon. This is stimulated emission triggered by the irradiated infrared light. The emitted light propagates away from the crystal to the outside, which means that part of the Joule energy to be used for diffusing B atoms is dissipated in the form of optical energy, resulting in local cooling that decreases the diffusion rate. As a result, by the balance between the heating by the Joule energy and the cooling by the stimulated emission, the spatial distribution of B atoms varies and reaches a stationary state autonomously.

It is expected that this DPP-assisted annealing will form the optimum spatial distribution of B atoms for efficient generation of DPPs, resulting in efficient device operation for light emission. Figure 1.6 shows the temporal evolution of the temperature of the device surface as the DPP-assisted annealing progressed. After the temperature rapidly rose, it fell and asymptotically approached a constant value. The features of this temporal evolution are consistent with those of the principle of the DPP-assisted annealing described above.

Figure 1.7a shows a photograph of a prototype Si light-emitting diode (LED) fabricated by this DPP-assisted annealing. It had an area as large as 12 $mm^2$. By injecting a forward current, the device emitted infrared light with a wavelength of 1.3 μm (Fig. 1.7b). The emitted optical power and the external quantum efficiency were as high as 1 W and 15% (at wavelengths in the range $1.32 \pm 0.15$ μm), respectively, at room temperature. Figure 1.7c and d respectively show the configuration and a photograph of a recently fabricated high-power device. Its area was as small as 1 $mm^2$ [33]. The output power was as high 200 mW, which means that the areal power density was three times that of the device in Fig. 1.7a, b.

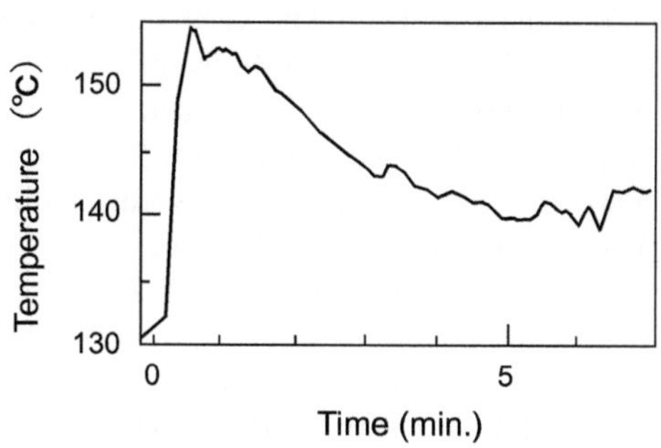


**Fig. 1.6** Temporal evolution of the temperature of the device surface as the DPP-assisted annealing progressed

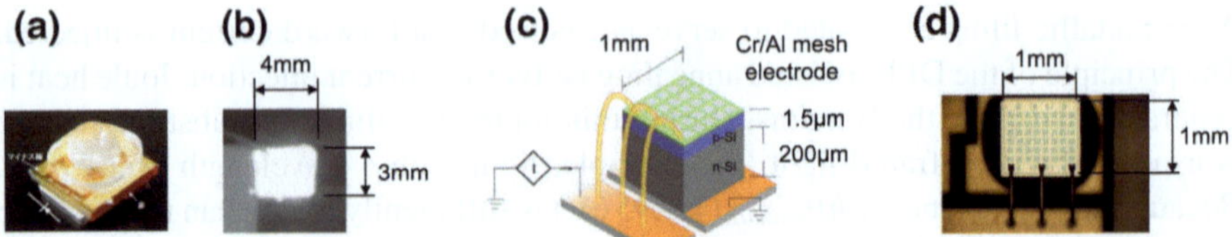


**Fig. 1.7** Photograph of prototype Si-LED. **a** External appearance of the fabricated device after packaging. **b** Emitted light profile. **c** and **d** show the configuration and a photograph, respectively, of the high-power device

**Fig. 1.8** Light emission spectra. Curves A–C are the spectra of the devices fabricated by DPP-assisted annealing for 1, 7, and 30 min, respectively

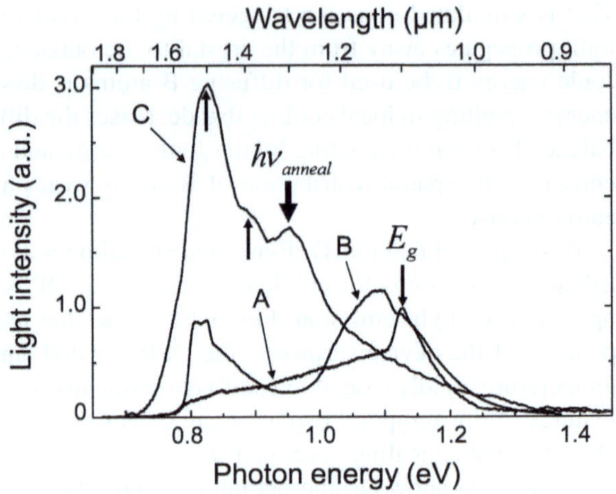


Figure 1.8 shows the light emission spectra of the fabricated Si-LED. Curves A-C are the spectra of devices fabricated by DPP-assisted annealing for 1, 7, and 30 min, respectively. They clearly show that the emitted light intensity increases with increasing DPP-assisted annealing time. The essential feature of these curves is the differences in their profiles: Although curve A has a peak around $E_g$, curve B shows a new peak at around 0.83 eV. In the case of curve C, no peaks were seen around $E_g$. Instead, a new peak appeared, identified by a downward thick arrow, at an energy that corresponds to the photon energy $h\nu_{anneal}$ of the light irradiated in the DPP-assisted annealing process. This peak is evidence that DPPs were created by the light irradiation, and that the B diffusion was controlled. Other evidence is that the photon energy of the emitted light, $h\nu_{em}$, was identical to that of the irradiated light, $h\nu_{anneal}$. That is, the irradiated light served as a breeder that created a photon with energy $h\nu_{em} = h\nu_{anneal}$. For this reason, this phenomenon is named photon breeding (PB) with respect to photon energy.

Here, the separations between the energies identified by two upward thin arrows (0.83 and 0.89 eV) on curve C, and by the down ward thick arrow (0.95 eV) were 0.06 eV, which is equal to the energy of an optical phonon in Si. This means that the two upward thin arrows show that the DPP with an energy of 0.95 eV was converted

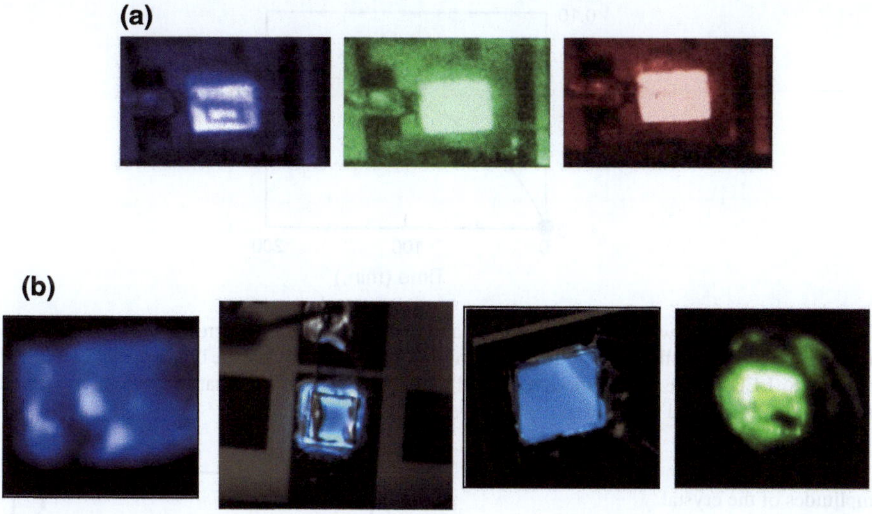

**Fig. 1.9** Light emitted from a variety of LEDs emitting visible light. **a** (Left to right) Blue, green, and red light emitted from Si-LEDs. **b** (Left to right) UV-violet, bluish-white, blue, and green light emitted from SiC-LEDs

to a free photon after emitting one and two optical phonons. This conversion process demonstrates that the light emission described here used the phonon energy levels as an intermediate state.

By using the novel PB phenomenon, a variety of LEDs have been fabricated by using crystalline Si. Specifically, blue, green, and red light emitting LEDs were fabricated by irradiating blue, green, and red light, respectively, during the DPP-assisted annealing (Fig. 1.9a) [34]. In order to increase the efficiency of extracting this visible light, a lateral p-n homojunction structure was developed [35].

Crystalline SiC is also a known typical indirect transition-type semiconductor. A variety of visible LEDs have been fabricated even using this material. They were fabricated by irradiating them with UV-violet, bluish-white, blue, and green light during the DPP-assisted annealing, and respectively emitted UV-violet, bluish-white, blue, and green light (Fig. 1.9b) [36–39].

PB was observed not only with respect to photon energy but also with respect to photon spin. That is, the polarization of the emitted light was equivalent to that of the light irradiated during the DPP-assisted annealing (Fig. 1.10) [40].

The origin of the PB was attributed to the spatial distribution of B atoms, which was controlled autonomously during the DPP-assisted annealing. By analyzing the three-dimensional spatial distribution of B atoms at the p-n homojunction, acquired by atom probe field ion microscopy with sub-nanometer resolution, it was found that the B atoms were apt to orient along a plane parallel to the top surface of the Si crystal and to form pairs with a length $d = 3a$, where $a$ (=0.54 nm) is the lattice constant of the Si crystal.

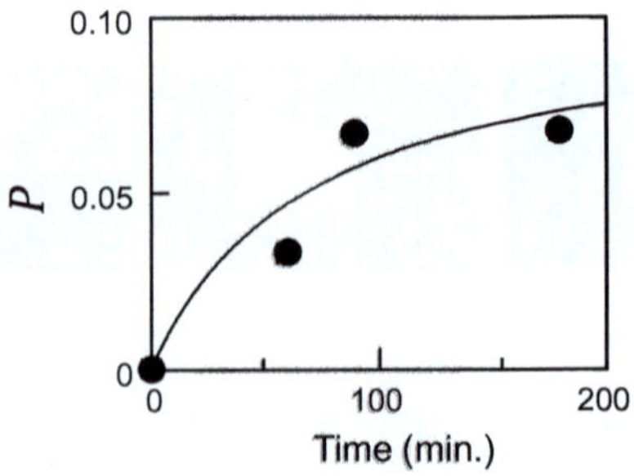


**Fig. 1.10** Relation between the DPP-assisted annealing time and the degree of polarization $P = (I_{\parallel} - I_{\perp})/(I_{\parallel} + I_{\perp})$ of the light emitted from the Si-LED. $I_{\parallel}$ and $I_{\perp}$ are the light intensities emitted from the Si-LED whose polarizations are parallel and perpendicular to that of the light irradiated during the DPP-assisted annealing, respectively

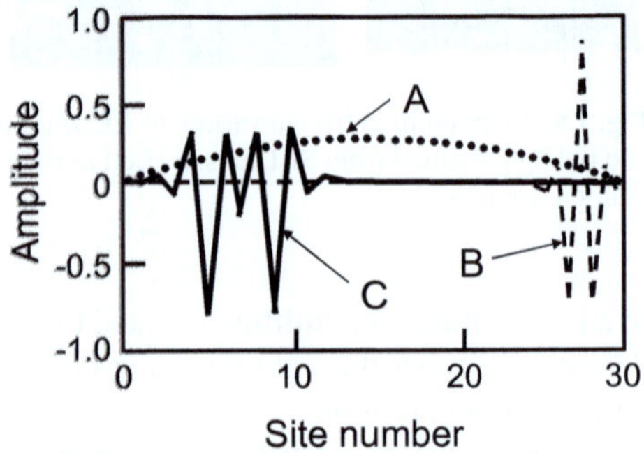


**Fig. 1.11** Vibration amplitudes of the crystal lattice of Si atoms. The number of lattice sites is 30. Curve A represents the non-localized mode. Curves B and C represent the first and second localized modes, respectively, where impurity atoms were assumed to be at sites 5, 9, 18, 25, 26, and 27

As a preliminary discussion on this origin, Fig. 1.11 shows the calculated vibration amplitudes of the crystal lattice of Si atoms, where a one-dimensional lattice was assumed for simplicity [30]. The curve A shows the amplitude of the lattice vibration in the case where the lattice is formed only by the Si atoms, which corresponds to a non-localized phonon mode. On the other hand, when some of the Si atoms are replaced by impurity B atoms, the amplitude profile of the lattice vibration changes greatly. As represented by curves B and C, the vibration amplitude is localized at the B atom-pair, because the B atom-pair serves as a cavity resonator to confine the lattice vibration. The confined lattice vibration corresponds to the localized phonon mode. As a result, impurity B atom-pairs serve as phonon localization centers, at which the DPPs can be created and localized efficiently.

The main discussion below follows from the preliminary discussion above: If $d$ is equal to the lattice constant $a$, the B atom-pair can orient in a direction parallel to the [1, 0, 0], [0, 1, 0], or [0, 0, 1] orientation because the Si crystal is composed of multiple cubic lattices. As a result, the momentum of the localized phonon points in this direction, which corresponds to the $\Gamma - X$ direction in reciprocal space (Fig. 1.12a). Thus, a photon is efficiently emitted because this $\Gamma - X$ direction is the same as the direction of the momentum of the phonon required for recombination between

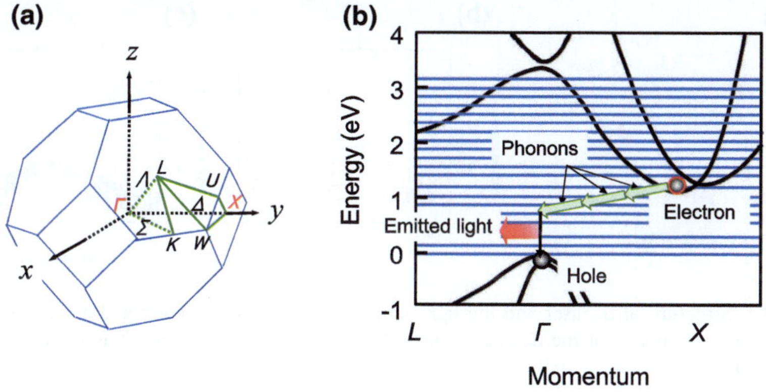


**Fig. 1.12** Energy band structure of Si. **a** A unit cell of the Si crystal in reciprocal lattice space. **b** Energy band structure and schematic illustration of light emission. Blue horizontal lines represent phonon energy levels involved in the light emission

an electron at the bottom of the conduction band at the $X$-point and a hole at the top of the valence band at the $\Gamma$-point. Here, it should be noted that the value of the momentum of the phonon has to be $h/a$ for this electron-hole recombination to take place. Furthermore, it should also be noted that the value of the momentum of the mode localized at the B atom-pair with $d = 3a$ is $h/3a$. By comparing these two values, it is found that the DPP at this B atom-pair has to create three phonons for recombination. In the other words, the B atom-pairs with $d = 3a$ most efficiently create three phonons for light emission. As a result, as is schematically shown in Fig. 1.12b, the emitted photon energy $h\nu_{em}$ is expressed as $h\nu_{em} = E_g - 3E_{phonon}$. By substituting the values of $E_g$ (=1.12 eV) and the relevant optical mode phonon energy $E_{phonon}$ (=65 meV) into this equation, the value of $h\nu_{em}$ is derived to be 0.93 eV, which is nearly equal to the photon energy $h\nu_{anneal}$ (=0.95 eV) irradiated during the DPP-assisted annealing. This numerical relation confirms that PB with respect to photon energy occurs.

In the case where the Si crystal surface is irradiated with linearly polarized light during the DPP-assisted annealing, analyses of the spatial distribution of B atoms confirmed that B atom-pairs with $d = 3a$ were also apt to be formed, and the direction of the B atom-pairs was normal to the polarization direction of the irradiated light. A possible origin of the induced polarization of the emitted light is as follows: First, when the Si-LED is fabricated by the DPP-assisted annealing, transverse optical phonons are created at the B atom-pairs and couple with the DPs. The vibration direction of these phonons is parallel to that of the electric field of the polarized light irradiated during the DPP-assisted annealing. Next, when the fabricated LED is operated, since these phonons are created again, the direction of the electric field vector of the emitted light becomes also parallel to the vibration direction of these phonons. Therefore, the polarization direction of the emitted light becomes identical to that of the light irradiated during the DPP-assisted annealing.

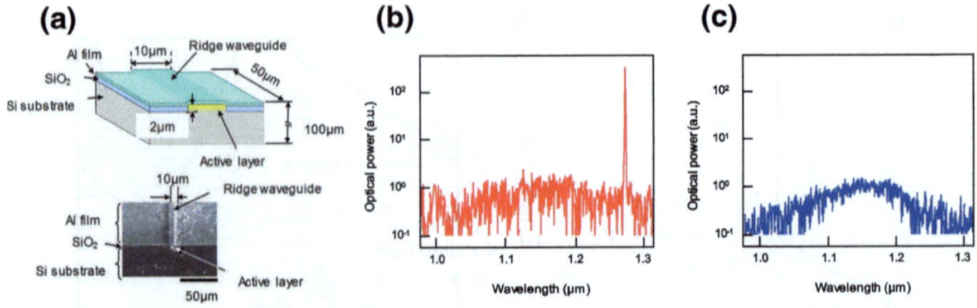


**Fig. 1.13** Structure of Si laser and the light emission spectra. **a** Structure and a scanning electron microscopic image of the device. **b** and **c** Spectral profiles above and below the threshold, respectively

In summary, the spatial distribution of B atoms was controlled autonomously by the DPP-assisted annealing to satisfy the momentum conservation law and to realize PB with respect to photon energy and to photon spin.

Si-lasers were also fabricated by using the DPP-assisted annealing. Figure 1.13a shows the device structure and a scanning electron microscopic image of the fabricated device [41]. A simple ridge waveguide was incorporated into the structure, and the cleaved facets were used as mirrors of a Fabry–Perot cavity. DPP-assisted annealing was then carried out by injecting 1.3 μm-wavelength light into the cavity through one of the end facets. Figure 1.13b, c show the light emission spectra of the fabricated Si-laser. Above the threshold, a sharp lasing spectrum was observed (Fig. 1.13b), which demonstrates single-mode oscillation at room temperature even though the cavity length was as long as 550 μm. The origin of this single-mode oscillation is that the low infrared absorption by the Si provides a low threshold for the principal longitudinal mode at the optical amplification gain spectral peak and, as a result, the gains for other modes are depleted by this principal mode due to nonlinear mode competition [42, 43]. The spectral profile below the threshold (Fig. 1.13c) does not show any ASE spectra, which is evidence of the gain depletion due to the mode competition above.

By modifying the device structure in Fig. 1.13a, a high-power infrared laser device was successfully fabricated by utilizing the very low infrared absorption of crystalline Si. Figure 1.14a shows a photograph of the device [44]. A cross-sectional profile is also shown, in which the ridge waveguide was not built-in because very efficient optical confinement was not expected by this waveguide as long as the device had a p-n homojunction. Instead, the cavity length was increased to 15 mm to realize high power. After the DPP-assisted annealing, the relation between the injected current density $J$ and the output power $P_{out}$ of the fabricated laser device was measured. As shown in Fig. 1.14b, an output power as high as 13 W was obtained. This value was more than $10^3$-times that of a conventional double heterojunction-structured InGaAsP/InP laser (10 mW at 1.3 μm wavelength: SLT1130 series manufactured by Sumitomo Electric). The threshold current density $J_{\mathrm{th}}$ was as low as 60 A/cm$^2$.

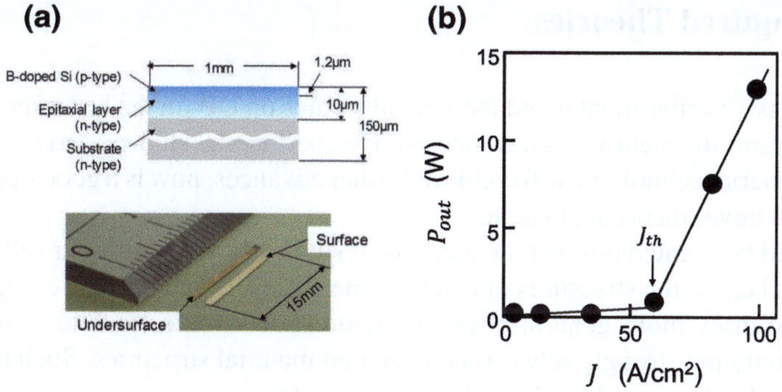


**Fig. 1.14** Structure of a high-power Si laser and the output power. **a** Structure and a photograph of the device. **b** Relation between the injected current density and the output optical power

Since crystalline Si was used without a built-in waveguide, the structure of the present device is more similar to those of solid-state and gas lasers than those of conventional double heterojunction-structured semiconductor lasers [45, 46]. Further similarities can be found by referring to the magnitude of the absorption loss: In the case of solid-state and gas lasers, electronic transitions in electronically isolated ions, atoms, and molecules are used for lasing. Therefore, even though direct electric current injection to these laser media is difficult, the absorption loss per unit volume is very low. Thus, a high optical output power can be obtained by increasing the size of the laser medium even though the photon density of the lasing light was low.

Since the DPPs used in the present Si laser were electronically isolated, as in the case of ions in the solid-state and gas lasers above, the absorption loss per unit volume was maintained very low. Thus, the threshold current density was very low. Furthermore, the optical output power was greatly increased by increasing the size of the Si crystal even though the photon density of the lasing light was low. An additional advantage was that electrons could be easily supplied by direct electric current injection, due to the low recombination loss in the crystalline Si.

For comparison, in the case of conventional semiconductor lasers, coupled electrons and holes in the conduction and valence bands, respectively, have been used for lasing. Therefore, direct electric current injection to the laser medium is easy. An additional advantage is that the laser medium can be very small. However, the problem was that the absorption loss per unit volume was large, making it impossible to achieve high optical output power and low threshold current density even though the photon density of the lasing light was high.

It has been believed for a long time that indirect transition-type semiconductors are unsuitable for use as laser media. Instead, direct transition-type semiconductors have been widely used until now [47]. However, thanks to the advent of DP science and technology, DPPs have resulted in the manifestation of large optical amplification gain in indirect transition-type semiconductors, which was the secret to the dramatically high optical output power and low threshold current density realized by using crystalline Si.

## 1.4 Required Theories

After extensive experimental and theoretical studies on DPs in the last three decades, the main effort in recent years has focused on experimental studies, and the development of generic technologies. To achieve further advances, now is a good opportunity to promote novel theoretical studies.

It should be pointed out that the theoretical studies carried out so far relied on NP structures, i.e., on the structures of electron energy levels and phonon energy levels. In order to draw more generic physical pictures, new routes to future theoretical studies should not strongly rely on such detailed material structures. Such routes are expected to be developed by the following three steps:

**Step 1**: To draw a more generic and basic physical picture of the DP.
**Step 2**: To draw a picture that allows us to treat multiple DPs created in a macroscopic material.
**Step 3**: To draw a picture for demonstrating how to take out the DP from a nano-system to a macro-system.

A hint to achieving step 1 can be found in the photon dispersion relation, i.e., the relation between the momentum and energy of a photon, as shown in Fig. 1.15. Since conventional optics deals with propagating light (free photons with a definite electromagnetic mode), its dispersion relation is represented by the black curve in this figure when the photon exists in a macroscopic material. The three-dimensional profile of this curve is a shell. The profile in vacuum is represented by the blue line, whose three-dimensional profile is a so-called light cone, which is a special case of the shell. That is, a free photon exists on the shell, and because of this, conventional optics can be named *on-shell science.*

In the casee case of a DP, on the contrary, it should be noted that its electromagnetic mode cannot be defined because of its nanometric nature, which makes the use of the dispersion relation invalid. This means that the DP does not exist on the shell.

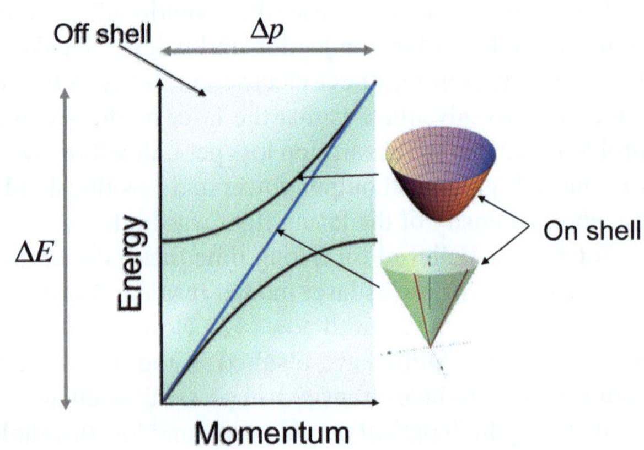


**Fig. 1.15** Dispersion relation. Two black curves represent the relation of a photon in a macroscopic material. The blue line is for a photon in free space. The green shaded rectangle represents that for the DP. The brown and green three-dimensional forms represent a shell and a light cone, respectively

Instead, it exists off the shell, a situation that is represented by the shaded green rectangle in this figure. Intrinsic features of the off-shell photon are: Since the momentum uncertainty $\Delta p$, represented by the horizontal double-headed arrow in this figure, is large, the size of the field $\Delta x$ is small, which is a consequence of the uncertainty principle $\Delta p \cdot \Delta x \geq \hbar$. Becau this feature, the DP has been called an optical near field. Furthermore, since the energy uncertainty $\Delta E$, represented by the vertical double-headed arrow in this figure, is also large, the duration of the created photon $\Delta t$ is short, which is also a consequence of the uncertainty principle $\Delta E \cdot \Delta t \geq \hbar$. A photon with this feature has been called a virtual photon. From these two intrinsic features, it is found that the off-shell photon is an optical near field and a virtual photon, which is the physical picture of the DP.

With the help of this hint, realization of step 1 was promoted by novel theoretical analysis of the Clebsch-dual electromagnetic field, for which a notion of a spacetime vortex field was used [48, 49]. The main results of this analysis are: (1) It was found that the DP exists in a spacelike domain of the Riemannian manifold. (2) The spatial profile of the DP is represented by the Yukawa function, which is equivalent to (1.8). Furthermore, it was found that the effective mass $m_{eff}$ of the DP field, being inversely proportional to the size $a$ of localization (a consequence of the Klein–Gordon's equation), is expressed as $m_{eff} = \sqrt{E^2 - \chi^2}$. Here, $E$ and $\chi$ are the instantaneous energy spent to create the DP and a characteristic scale of the NP, respectively. It suggests that the DP can be transferred from the spacelike to the timelike domain if $E > \chi$, and that the DP can be transformed to be observable, however, within the very short duration $\Delta t$ above. (3) It was found that the energy-momentum tensor of the DP became isomorphic to Einstein's equation, which implies that the DP is closely related to the vacuum energy.

The reasons why step 2 is indispensable are: As was reviewed in the previous section describing the DPP-assisted annealing of crystalline Si, there were too many Si atoms, electrons, B atoms, DPs, and DPPs in the crystal. Since it is difficult to treat these multiple quasi-particles and elementary particles by the conventional deterministic method of theoretical analysis, a method of avoiding this difficulty was proposed by considering a system composed of these mutually interacting particles as a complex system [50].

Based on this consideration, computer simulations were carried out by using a stochastic model. Here, an interactive random-walk process on a crystal lattice and phenomenological coupled-Poisson process were assumed to describe the diffusive motion of doped B atoms and electron-phonon coupling, respectively. As a result, the main experimental results were successfully reproduced, such as the temporal behaviors of the crystal temperature and the emitted light intensity observed during the DPP-assisted annealing.

Step 3 is indispensable, especially for connecting the theoretical and experimental studies. In the experiment, it is essential to create or detect the DP in the most efficient manner; that is to say, the most efficient micro-macro conversion is required. To meet this requirement, high-quality fiber probe tips (Fig. 1.1) have been developed [6] in order to create or detect DPs with high spatial resolution or high sensitivity. The concept of micro-macro duality in the quantum field has been found to be a powerful

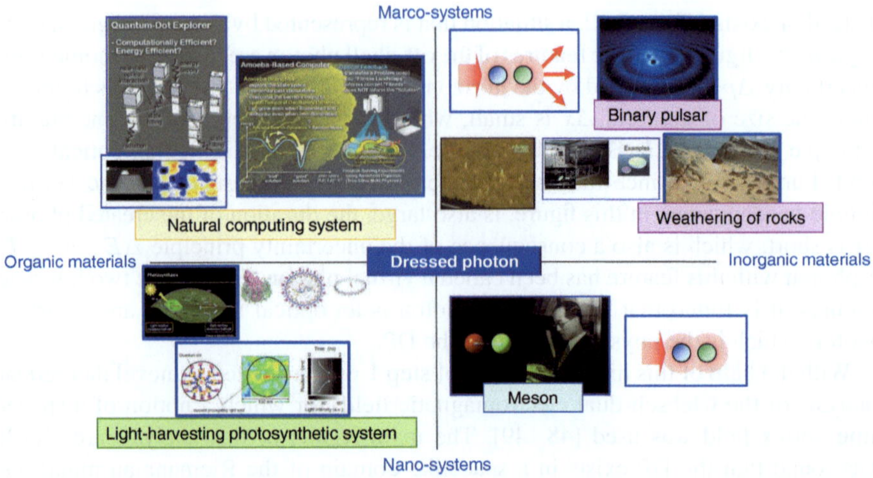


**Fig. 1.16** A variety of phenomena occurring in nature that have similar features to those of the dressed photon

and promising theoretical tool for drawing a picture of the micro-macro conversion [51], and detailed studies have commenced [52].

In order to support further progress in achieving steps 1–3 above, another route to future theoretical studies is also under development, based on the classical electromagnetic field theory [53, 54]. These studies recently found that the longitudinal electric field, which has been ignored in conventional optics, induced a non-resonant light-matter interaction in a nanometric space, as has been previously observed in several experiments [55].

Figure 1.16 represents a variety of phenomena occurring in nature, which have similar features to those of the DP. They have been found in nano-systems and in macro-systems, as well as in organic materials and in inorganic materials. In an inorganic nano-system, a representative example is a meson, which connects two nucleons. Some features of the meson are similar to those of the DP, which connects NPs by means of DP energy exchange. It has been known that the potential profile of the meson is expressed by the Yukawa function, shown in (1.8).

In an organic nano-system, an example is a light-harvesting photosynthetic system, whose light trapping operation is similar to the operation of logic gate devices based on DP energy transfer [56]. In an organic macro-system, an example is the natural computing observed in single-celled amoeboid organisms [57], whose signal transmission features are similar to those of DP energy transfer. Some features of this computing have been demonstrated by using the logic gate devices described above [58].

Finally, one example of an inorganic macro-system is the photochemical reaction in the weathering of rocks. Some features of this reaction are similar to the photochemical reaction induced by a DP, which has been used for polishing material surfaces to atomic-level flatness [59]. Another example is a binary pulsar [60]. Some of the features of the gravitational wave radiation from a binary pulsar are similar to the propagating light radiation from two NPs as a result of DP energy exchange and relaxation.

By referring to Fig. 1.16, it can be understood that the DP is not a special topic of a narrow field of science but is connected to more general and broader scientific fields.

## 1.5 Toward Steps 1 and 2

This section presents reasons why steps 1 and 2 in Sect. 1.4 are required for drawing a generic physical picture of the DP. For this end, unique features of DP energy transfer between NPs are reviewed. Following this review, Sect. 1.6 introduces the principles and practices of novel functional devices and an optical energy converter. Section 1.7 reviews the application of this DP energy transfer to optical energy conversion. In Sects. 1.6 and 1.7, spatial and temporal evolutions of the DP energy transfer are demonstrated.

Novel functional devices that operate by using DPs are named DP devices. Some of these devices are introduced in Sect. 1.6. In order to examine the principle of operation of DP devices, two different-sized cubic semiconductor NPs are assumed, in which exciton energy is quantized, as schematically explained by Fig. 1.17. Small and large NPs ($NP_S$ and $NP_L$, respectively) in this figure are used as input and output terminals, respectively. These NPs are irradiated with propagating light whose photon energy is resonant with the lowest quantized energy level (1, 1, 1) of the exciton in $NP_S$. This light corresponds to the input signal to the DP device. The propagating light emitted from the lowest energy level (1, 1, 1) in the large $NP_L$ corresponds to the output signal.

Based on the localized nature of the DP (1.8), these NPs are installed in close proximity to each other, with a separation as short as their sizes. By assuming that the ratio between their side lengths is $1 : \sqrt{2}$, one can readily find that the exciton energy of level (1, 1, 1) in $NP_S$ is equal to that of level (2, 1, 1) in $NP_L$; that is, these two energy levels are resonant with each other. It should be noted that level (2, 1, 1) in $NP_L$ is an electric-dipole forbidden level, whereas level (1, 1, 1) in $NP_S$ is

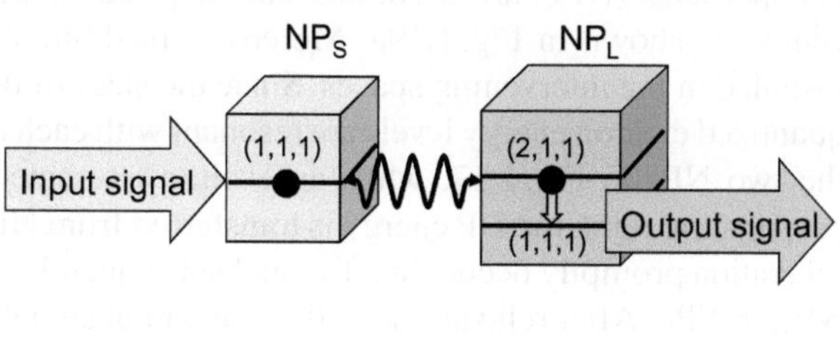


**Fig. 1.17** DP energy transfer and dissipation in two cubic semiconductor NPs

an electric-dipole allowed level. Thus, the energy of the irradiated propagating light normally cannot be transferred from level (1, 1, 1) in $NP_S$ to level (2, 1, 1) in $NP_L$.

However, because of the localized nature of the DP, there is a way for the energy to be transferred from $NP_S$ to $NP_L$, as represented by the double-headed wavy arrow in Fig. 1.17. The transfer process is:

(1) An exciton is excited to level (1, 1, 1) in $NP_S$ by the irradiated propagating light.
(2) A DP is created by this exciton and is localized on $NP_S$.
(3) The energy of this DP is transferred to $NP_L$.
(4) An exciton is excited to level (2, 1, 1) in $NP_L$ even though this level is electric-dipole forbidden.

After this process, the exciton promptly relaxes to the lower level (1, 1, 1) in the $NP_L$ and dissipates a small amount of energy, as shown by the downward arrow. Then, the relaxed exciton emits propagating light, to be used as the output signal.

## 1.6 Application to Functional Devices

The unique transfer and dissipation of the DP energy, reviewed in the previous section, have been applied to the development a variety of DP devices, such as AND and NOT logic gates, by integrating size-controlled semiconductor NPs on a substrate. Figure 1.18 demonstrates the structure and operation of these devices [61]. In addition to these logic gate devices, a variety of other DP devices have been developed, including a nano-optical condenser [62], a digital-to-analog converter [63], an energy transmitter [64], a frequency up-converter [65], a delayed-feedback optical pulse generator [66], and so on. As representative examples of these devices, this section reviews a nano-optical condenser and an energy transmitter.

### *1.6.1 Nano-Optical Condenser*

A nano-optical condenser that converts propagating light to DPs with high efficiency has been developed on the basis on the unique spatial evolution features of the DP energy transfer [62]. In order to construct this device, a large number of small nanoparticles ($NP_S$) are used, and one large nanoparticle ($NP_L$) is installed at the center, as shown in Fig. 1.19a. Moreover, medium-sized nanoparticles ($NP_M$) are installed in the intervening spaces. Since the sizes of these NPs are tuned so that the quantized exciton energy levels are resonant with each other, as was the case between the two NPs in Fig. 1.17, when an exciton is created in $NP_S$ by irradiation with propagating light, the DP energy is transferred from $NP_S$ to $NP_M$. After this transfer, relaxation promptly occurs in $NP_M$, and subsequently, the energy is transferred from $NP_M$ to $NP_L$. After relaxation in $NP_L$, the output signal is generated from the exciton in the lowest energy level. Here, since the energy level (1, 1, 1) in $NP_S$ is tuned to the

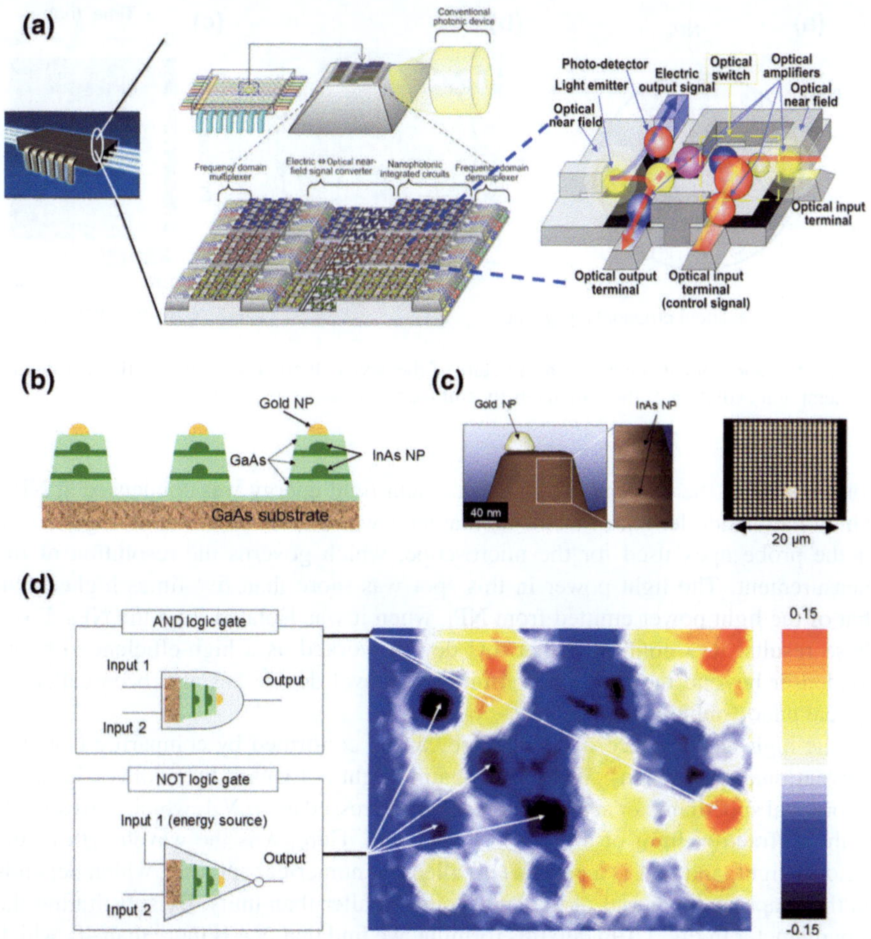


**Fig. 1.18** Dressed photon devices. **a** Prototype of integrated circuit composed of DP devices. **b** Cross-sectional structure of mesa-shaped NOT logic gate composed of InAs NPs. **c** Scanning transmission electron microscope image of structure in **b** (left) and optical microscope image of a two-dimensional array of structures in **b** (right). **d** Measured spatial distribution of the output signal intensity from a two-dimensional array of fabricated devices composed of InAs NPs

photon energy of the incident propagating light, almost all the incident propagating light energy can be absorbed by a large number of NPs.

Furthermore, since the energy dissipation in this system is due to the relaxation in $NP_M$ and $NP_L$, its magnitude is very small. As a result, a high efficiency is expected for concentrating the energy of the incident propagating light at $NP_L$.

Figure 1.19b shows an observed near field optical microscopic image of the spatial distribution of the DP energy by applying propagating light with a wavelength of 385 nm to cubic CuCl NPs in an NaCl host crystal. The bright spot at the center

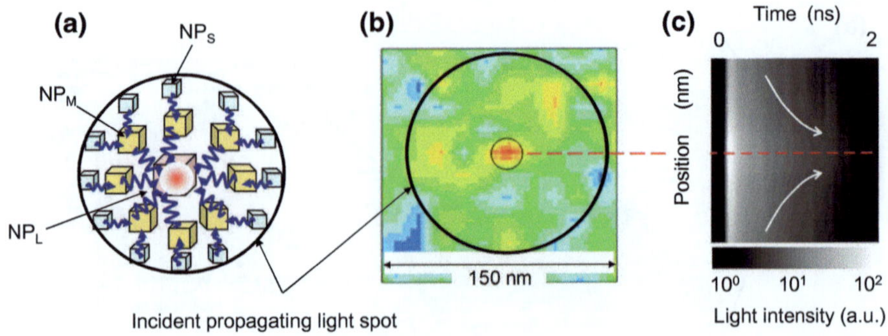


**Fig. 1.19** Nano-optical condenser. **a** Structure of the device. **b** and **c** show the spatial distribution and temporal evolution of the light intensity emitted from CuCl NPs, respectively

corresponds to the place at which the incident light energy was condensed at $NP_L$, which had a side length of 8 nm. Its diameter was about 20 nm, including the size of the probe apex used for the microscope, which governs the resolution of the measurement. The light power in this spot was more than five-times higher than that of the light power emitted from $NP_L$ when it was isolated from the NPs. From these results, it is confirmed that this device worked as a high-efficiency optical condenser beyond the diffraction limit. This novel device has also been called an optical nano-fountain [62].

The high performance of this device can be confirmed by comparing it with a conventional convex lens. When propagating light is focused by a convex lens, the theoretical spot diameter at the focal plane is expressed as $\lambda/NA$, which corresponds to the diffraction limit of a focused light beam. Here, $\lambda$ is the wavelength of the incident light, and $NA$ is a parameter called the numerical aperture, which depends on the shape and material of the lens, being smaller than unity. By substituting the spot diameter in Fig. 1.19b into this formula, we find that $NA$ is more than 40, which is much larger than that of a conventional convex lens.

Figure 1.19c shows the measured spatial and temporal evolutions of the light intensity. The horizontal axis at the top of the figure represents time, and the vertical axis represents the radial position in polar coordinates centered at $NP_L$. The brightness gradation is proportional to the number of emitted photons, from which one can find that the energy is condensed at $NP_L$ with a time constant as short as 2 ns.

Because of the extremely low energy dissipation due to the relaxation from the upper to lower energy levels in $NP_M$ and $NP_L$, the efficiency of optical energy concentration can be higher than 0.9. The energy transfer process in the nano-optical condenser described above is similar to that in photosynthetic bacteria [67], whose high energy transfer efficiency is receiving attention as a novel system function inherent to such complex systems in a nano-scale space [68, 69].

### 1.6.2 Energy Transmitter

In addition to the nano-optical condenser described above, another example that uses the spatial evolution features of the DP energy transfer is an energy transmitter. This transmitter is used to transmit a signal from one DP device to another, corresponding to the function of a metallic wire in an electrical circuit or an optical waveguide in a conventional optical integrated circuit. It should meet the following two requirements:

(1) Signal reflection from the DP devices connected to the tail of this transmitter must be avoided to achieve stable uni-directional energy transmission.
(2) Transmission loss must be sufficiently low to realize a long transmission length.

Figure 1.20 shows the structure of an energy transmitter that was developed to meet these requirements. It is composed of an array of $N$ NPs of the same size ($NP_1$–$NP_N$), and a large NP ($NP_O$) installed at the end of this array. In the case of using cubic NPs, as an example, an exciton is created in energy level (1, 1, 1) in $NP_1$ by applying an input signal, i.e., by irradiating $NP_1$ with propagating light. This light is converted to a DP and is transferred to energy level (1, 1, 1) in $NP_N$, and nutation occurs among the $N$ NPs. As a result, these NPs are coupled with each other. If the size of $NP_O$ is tuned so that its electric dipole-forbidden level (2, 1, 1) is resonant with the energy level of this coupled state of the $N$ NPs, nutating energy is transferred to the upper energy level (2, 1, 1) in $NP_O$, and by subsequent relaxation to the lower energy level (1, 1, 1). Finally, the light is emitted from the exciton in level (1, 1, 1) and is used as the output signal.

The device in Fig. 1.20 meets requirement (1) above because the exciton cannot be excited to the upper energy level (2, 1, 1) in $NP_O$ even if the exciton is created in the lower energy level (1, 1, 1) in $NP_O$ by back-transfer of the signal from the DP devices installed at the stage after $NP_O$. Thus, the energy is not back-transferred from $NP_O$ to $NP_1$–$NP_N$.

It can be easily found that requirement (2) is met because the magnitude of the energy dissipated during the relaxation from level (2, 1, 1) to level (1, 1, 1) in $NP_O$ is as low as 20 meV, which is much lower than the photon energy of the light radiated onto the device.

For device fabrication, it is much easier to randomly disperse $NP_1$–$NP_N$ and $NP_O$ on a substrate than to arrange them accurately so as to maintain a constant separation. Figure 1.21a schematically explains this configuration [70], in which small NPs are

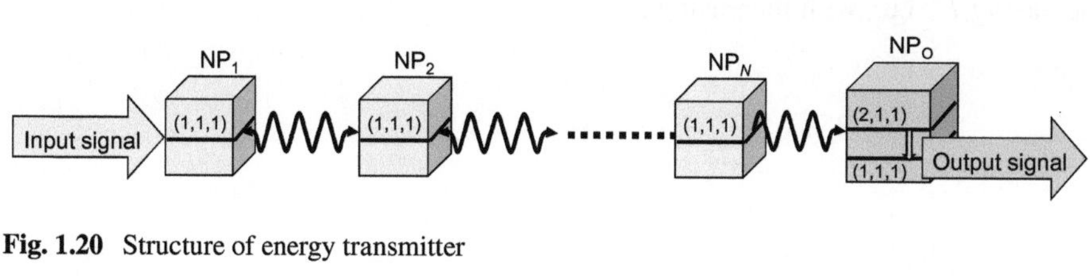


**Fig. 1.20** Structure of energy transmitter

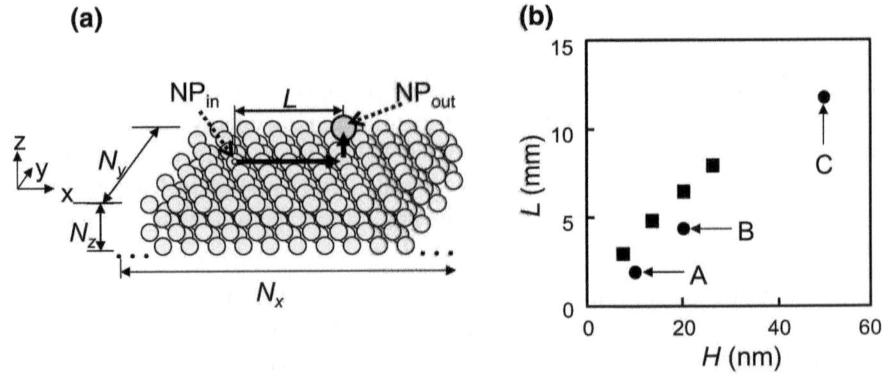


**Fig. 1.21** Calculated results for NPs dispersed on a planar substrate. **a** Arrangement of multiple small NPs ($NP_S$) and one larger NP ($NP_L$). **b** Dependence of the energy transfer length $L$ on the thickness $H$ of the small NP layers

randomly dispersed along the $x$-, $y$-, and $z$-axes, and are used as $NP_1$–$NP_N$, whose numbers of rows are denoted by $N_x$, $N_y$, and $N_z$, respectively. $NP_O$ is installed among the dispersed small NPs. $NP_1$ and $NP_O$ are respectively denoted by $NP_{in}$ and $NP_{out}$ in this figure.

For experiments, spherical CdSe NPs were used for simplicity of material preparation. The diameters of $NP_1$–$NP_N$ were maintained at 2.8 nm, whereas that of $NP_O$ was 4.1 nm in order to satisfy the resonance condition of the exciton energy levels. These spherical CdSe NPs were dispersed on a $SiO_2$ substrate, and the average separation between the adjacent NPs was arranged to be close to 7.3 nm in order allow efficient DP energy transfer (1.8). Moreover, the thickness of the NP layers, $H$ in Fig. 1.21a, was fixed to 10, 20, and 50 nm, which is proportional to the number of rows $N_z$ of $NP_1$–$NP_N$ along the $z$-axis. These devices are denoted by A, B, and C, respectively.

By applying propagating light with a wavelength of 473 nm, the energy transfer length $L$ was measured as a function of $H$. The results are shown in Fig. 1.21b, from which the values of $L$ for devices A, B, and C were found to be 1.92, 4.40, and 11.8 μm, respectively. These are much longer than the wavelength of the incident light, which also meets requirement (2) above. This figure shows that these measured values agree with the values calculated by using the rate equations representing DP energy transfer between two adjacent NPs. It also shows that $L$ increases with increasing $H$, i.e., with increasing $N_z$.

### 1.6.3 Autonomy in Dressed Photon Energy Transfer

The DP devices reviewed in the previous subsections exhibit novel characteristics which are superior to those of conventional photonic devices. They are: single-photon operation [71], low energy dissipation [72], low energy consumption [73], tamper resistance [74], and skew resistance [75]. Furthermore, an outstanding advantage conferred by the spatial evolution features is autonomy in the DP energy transfer [76], which is reviewed in this subsection.

As is schematically explained by Fig. 1.22a, the present model contains $N$ small NPs ($NP_S$) and one large NP ($NP_L$). By assuming that each $NP_S$ is initially occupied by an exciton, quantum master equations for the density matrix are solved to derive the occupation probability of the exciton in the lower energy level in $NP_L$. The time-integrated value of this probability corresponds to the output signal intensity.

This intensity is calculated as a function of the number $N$ of CuSe $NP_S$ The calculated results are indicated by closed circles in Fig. 1.22b and show that the efficiency in energy transfer is highest when $N \simeq 4$. Since the radiative relaxation rate from the lower energy level in $NP_L$ takes a finite value, the DP energy is not transferred to $NP_L$ until the exciton in the lower energy level is annihilated, and as a result, the energy is dissipated from $NP_S$ if $N$ is too large. Therefore, the output signal intensity does not increase if too many $NP_S$ are installed around an $NP_L$, and as a result, the efficiency of the energy transfer to $NP_L$ decreases when $N$ is larger than 4.

Small and large spherical CdSe NPs (2.0 and 2.8 nm diameters, respectively) were used, as was the case described in Sect. 1.6.2, to experimentally measure the magnitude of the energy transferred from $NP_S$ to $NP_L$. The results are represented by the closed squares in Fig. 1.22b, which show that the output signal intensity takes the maximum at $N \simeq 4$.

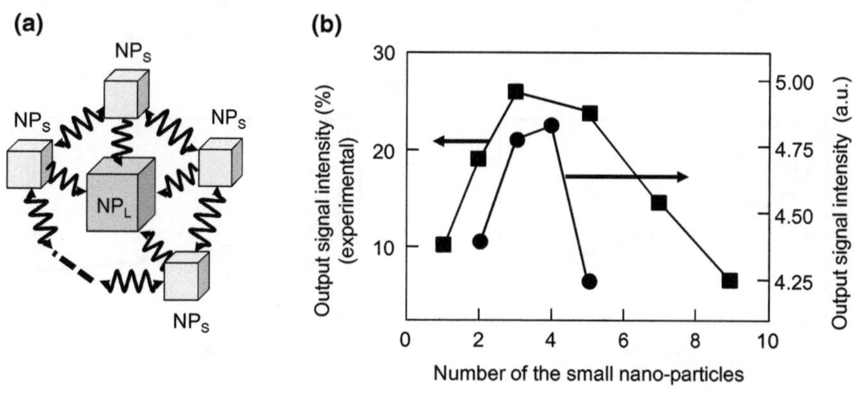


**Fig. 1.22** Autonomy of the DP energy transfer. **a** Layout of NPs. **b** Dependences of the output signal intensities emitted from CuSe NPs on the ratio of the number of $NP_S$s to that of $NP_L$

The dependence of the energy transfer on the number $N$ of the NPs suggests that the output signal intensity can be controlled by designing the positions of the NPs. Let us assume that interactions between some $NP_S$ and $NP_L$ may be degraded or lost because their resonant conditions are not satisfied due to, for example, size-detuning of NPs, fluctuations in the separations between NPs, and deterioration of the NP materials. In the case of a pentagonal layout, as shown in Fig. 1.23a, there can be eight degraded configurations: By referring to system E0 without any degradations, system E1 represents the layout in which the interaction between one $NP_S$ and $NP_L$ is degraded or lost (represented by the mark × between $NP_{S1}$ and $NP_L$ in this figure). Systems E2 and E2' have two degraded interactions.

Figure 1.23b shows the relation between the systems (E0–E5) and the time-integrated values of the occupation probability of the exciton in $NP_L$. This figure shows that system E5 does not generate any output signals because the interaction between $NP_S$ and $NP_L$ is completely lost. In contrast, the output signal intensities from systems (E1–E4) with degraded interactions are larger than that from system E0. In particular, the value of the output signal intensity from system E2 is 1.64-times that from system E0, which is consistent with the fact that the signal intensity takes the maximum at $N \simeq 4$ in Fig. 1.22b.

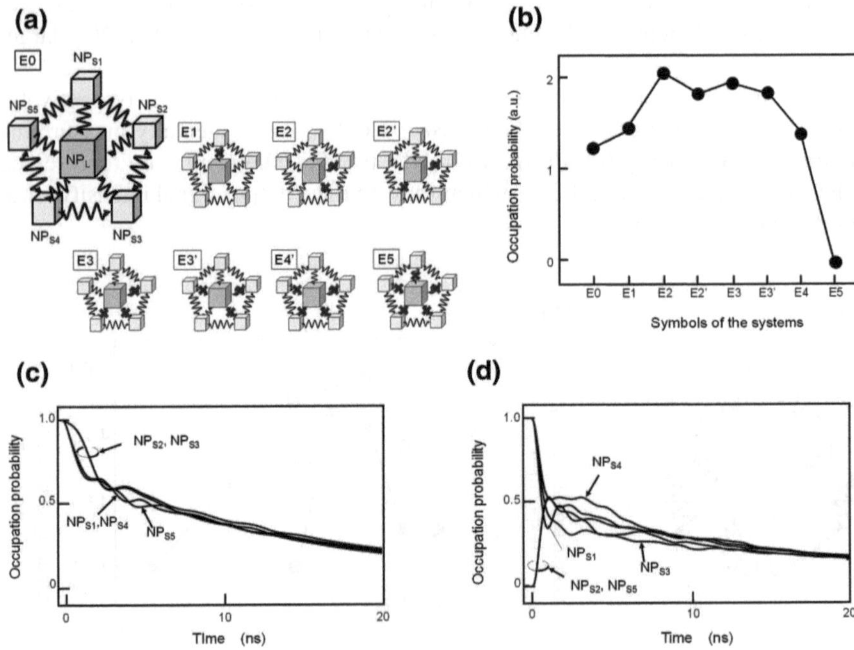


**Fig. 1.23** Degraded or lost interaction between $NP_S$s and $NP_L$. **a** Layout of NPs. **b** Time-integrated values of the occupation probabilities for systems E0 to E5. **c** and **d** are temporal evolutions of the occupation probabilities of the exciton in the energy levels in five $NP_S$s in systems E2 and E0, respectively

Moreover, the autonomy in energy transfer can be understood from Fig. 1.23c. This figure shows the temporal evolutions of the occupation probabilities of excitons in the energy levels in five $NP_S$ in system E2, in which two interactions are degraded ($NP_{S2}$–$NP_L$ and $NP_{S3}$–$NP_L$), as was shown in Fig. 1.23a. The energy levels in all the NPs are initially occupied by excitons, and afterward for several ns, the occupation probabilities in $NP_{S2}$ and $NP_{S3}$ remain high, which means that the energy is efficiently stored in $NP_{S2}$ and $NP_{S3}$ until it is transferred to $NP_L$. On the other hand, Fig. 1.23d shows the time evolutions of the occupation probabilities in the case of system E0, in which the energy levels in three $NP_S$ ($NP_{S1}$, $NP_{S3}$, and $NP_{S4}$) are initially occupied by excitons. It is found from this figure that the occupation probabilities for $NP_{S2}$ and $NP_{S5}$ increased within 2 ns even though they were initially zero. This means that, in a sense, the transferred energy autonomously searches for unoccupied $NP_S$s in the system.

It was demonstrated that a single energy transfer process is about $10^4$-times more efficient compared with the single bit-flip energy required in conventional electronic devices [72]. On the other hand, energy transfer in light harvesting antennas exhibits superior efficiency [77], and these structures have similarities to nanostructures networked via interactions by DP energy transfer. In summary, these studies will be extremely helpful for developing advanced DP devices with higher performance.

### *1.6.4 Temporal Evolution of Dressed Photon Energy Transfer*

In addition to the spatial evolutions shown in Sects. 1.6.2 and 1.6.3, Fig. 1.24a shows the temporal evolution of the output signal from the DP device, i.e., the light intensity from $NP_L$, emitted when a propagating light pulse (pulse width: 10 ps) is applied to $NP_S$. Black circles are the measured values acquired for $0 \le t \le 4$ ns. The optical intensity increases rapidly with a rise time $\tau_r$ of 90 ps, which depends on the magnitude of the transferred DP energy. After the applied signal pulse decays, the output signal also decays with small amplitude oscillation. This oscillation originates from the nutation of the DP energy transfer between $NP_S$ and $NP_L$. The period of oscillation is found to be about 400 ps from this figure. The decay time, i.e., the fall time $\tau_f$, is about 4 ns, which depends on the value of the radiative relaxation rate from the low energy level in $NP_L$. The solid curve represents the result calculated by using quantum master equations for the density matrix [78], which agrees well with the experimental results.

For more detailed analyses of the temporal evolution of the DP energy transfer, the optical intensity was acquired in the time range $0 \le t \le 10$ ns, which is longer than that in Fig. 1.24a. Black squares in Fig. 1.24b are the acquired values. The curve A represents the temporal evolution expressed as $\exp(-t/\tau_{f1})$, where $\tau_{f1}$ is the fall time, depending on the magnitude of the transferred DP energy. This curve agrees with the black squares only for an initial stage as short as $0 \le t \le 2$ ns. On the other hand, the curve B represents the temporal evolution expressed as $\exp(-\sqrt{t/\tau_{f2}})$, where the fall time $\tau_{f2}$ is the radiative relaxation rate from the low energy level

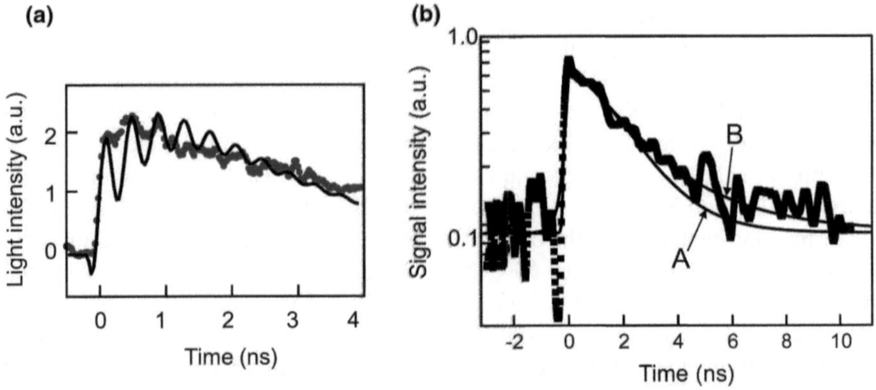


**Fig. 1.24** Temporal evolution of the light intensity emitted from the CuCl $NP_L$. **a** and **b** are the optical intensities, acquired for $0 \le t \le 4$ ns and for $0 \le t \le 10$ ns, respectively

in $NP_L$. This agrees with the black squares for a wide range of time periods up to 10 ns. For more detailed analyses of the DP energy transfer dynamics, it will be advantageous to suppress the component expressed as $\exp(-\sqrt{t/\tau_{f2}})$, which can be realized by decreasing the device temperature.

## 1.7 Application to Optical Energy Conversion

A novel thin film was invented recently, which efficiently converts ultraviolet (UV) photon energy to visible photon energy by means of DP energy transfer and dissipation. It has been used to drastically improve the electrical power generation efficiency of a solar cell battery. This section reviews the principle and operation of this film.

### *1.7.1 Principle*

Transparent silicone or ethylene-vinyl acetate (EVA) was used as the film material, with a thickness of 10 m. ZnO semiconductor NPs (average size and number density are 3–5 nm and $10^{17}$–$10^{18}$/cm$^3$, respectively) and DCM dye NPs (weight density 3–5 mg/cm$^3$) were dispersed in this film. It is required that the separation between the ZnO and DCM NPs be as close as possible to their sizes in order to transfer the DP energy between them (1.8). The number density and weight density above meet this requirement [79, 80].

The roles of the ZnO and DCM NPs in this film are to absorb UV light and emit visible light, respectively. They correspond respectively to $NP_S$ and $NP_L$ in Fig. 1.17. That is, UV light absorption excites an electron in the ZnO NP to create the DP.

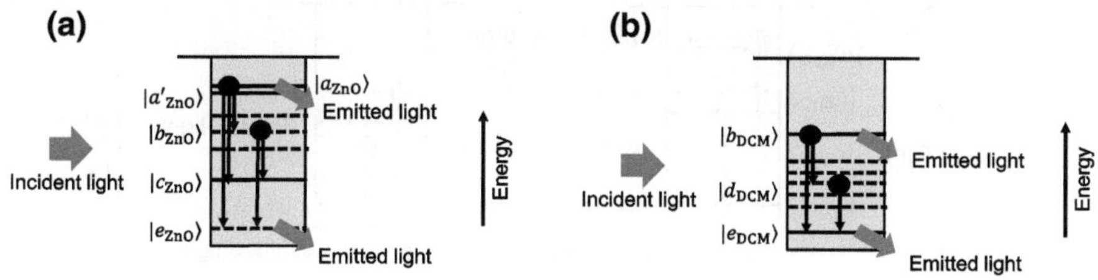


**Fig. 1.25** Electronic energy levels in a ZnO NP (**a**) and a DCM NP (**b**)

Then, the DP energy is transferred to the DCM NP, resulting in excitation of the electron in the DCM NP. This electron dissipates a small amount of its energy by relaxing to a lower energy level, emitting visible light.

For the DP energy transfer, an electronic dipole-forbidden transition is utilized (refer to the paragraph below (1.8)). Furthermore, since the magnitude of the energy dissipation above is very low, very high-efficiency optical energy conversion can be realized.

Figure 1.25a, b show the electronic energy levels in the ZnP and DCM NPs for UV light absorption and visible light emission, respectively. They are:

[In ZnO NP]

$|a_{ZnO}\rangle$ : An energy level of the electron in the conduction band. Since it is an electric dipole-allowed level, the electron is excited to this level by UV light absorption.
$|a'_{ZnO}\rangle$ : An electric dipole-allowed energy level in the conduction band. Its energy is slightly lower than that of $|a_{ZnO}\rangle$.
$|b_{ZnO}\rangle$ : Several energy levels in the bandgap, which originate from impurity atoms in the ZnO NP. They consist of electric dipole-allowed and -forbidden energy levels.
$|c_{ZnO}\rangle$ : An electric dipole-allowed level of the energy levels $|b_{ZnO}\rangle$ above, which is a defect level originating from the oxygen atoms in the ZnO NP. The electron in this level emits blue light.
$|e_{ZnO}\rangle$ : An electric dipole-forbidden level of the energy levels $|b_{ZnO}\rangle$ above. However, it can emit light because it is the lowest energy level of $|b_{ZnO}\rangle$.

[In DCM NP]

$|b_{DCM}\rangle$ : An electric dipole-allowed level. Thus, the electron in this level can emit light. It is resonant with the energy level $|b_{ZnO}\rangle$ in the ZnO NP.
$|d_{DCM}\rangle$ : Electric dipole-forbidden energy levels, which originate from the impurity atoms in the DCM NP. Some of them are resonant with the energy level $|c_{ZnO}\rangle$ in the ZnO NP.
$|e_{DCM}\rangle$ : An electric dipole-allowed level of the energy levels $|d_{DCM}\rangle$. Thus, the electron in this level can emit light. It is resonant with the energy level $|e_{ZnO}\rangle$ in the ZnO NP.

The UV light absorption and visible light emission processes in the ZnO NP and DCM NP, are:

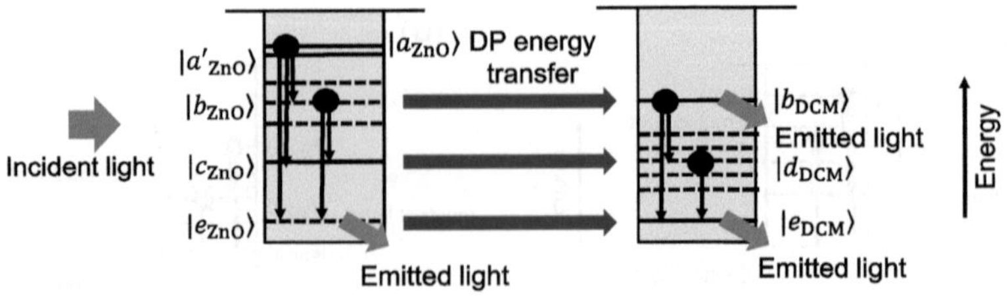


**Fig. 1.26** Light absorption and emission of a film in which ZnO and DCM NPs are dispersed

[In ZnO NP (Fig. 1.25a)]

An electron in the ZnO NP is excited to the energy level $|a_{ZnO}\rangle$ by UV light absorption. This excited electron has a higher probability of UV light emission than that of relaxation to a lower energy level. Thus, the conversion efficiency from UV light to visible light is low.

[In DCM NP (Fig. 1.25b)]

After an electron in the DCM NP is excited to a higher energy level by UV light absorption, it relaxes to an electric dipole-forbidden energy level, which is a triplet energy level. Thus, the conversion efficiency from UV light to visible light is low.

However, in the case where both the ZnO and DCM NPs are dispersed in the film, the light absorption and emission processes are remarkably different from those above. They are (Fig. 1.26): An electron in the ZnO NP is excited to the energy level $|a_{ZnO}\rangle$ by UV light absorption. It subsequently relaxes to the lower energy levels $|b_{ZnO}\rangle$, $|c_{ZnO}\rangle$, or $|e_{ZnO}\rangle$. Then, a DP is created, and its energy is transferred to the resonant energy levels $|b_{DCM}\rangle$, $|d_{DCM}\rangle$, or $|e_{DCM}\rangle$ in the DCM NP. As a result, an electron is excited to these levels and emits visible light. This visible light emission realizes a high efficiency of energy conversion from UV light to visible light. It should be noted that the DP energy can be transferred back from the DCM to the ZnO particles even though the electron in the DCM particles relaxes to the triplet energy level. This back transfer contributes to further increases in the energy conversion efficiency.

### 1.7.2 Operating Characteristics

Figure 1.27 shows measured excitation spectra, in which the horizontal axis represents the wavelength of the light incident on the film, and the vertical axis is the intensity of the visible light (wavelength 560 nm) emitted from the electron in the energy level $|e_{DCM}\rangle$ of the DCM NP. Figure 1.27a, b show the results acquired when only the ZnO and DCM NPs are dispersed in the film, respectively. Figure 1.27c shows those when both ZnO and DCM NPs are dispersed. Comparing the areas below the curves in these figures, the efficiency of the conversion from UV light

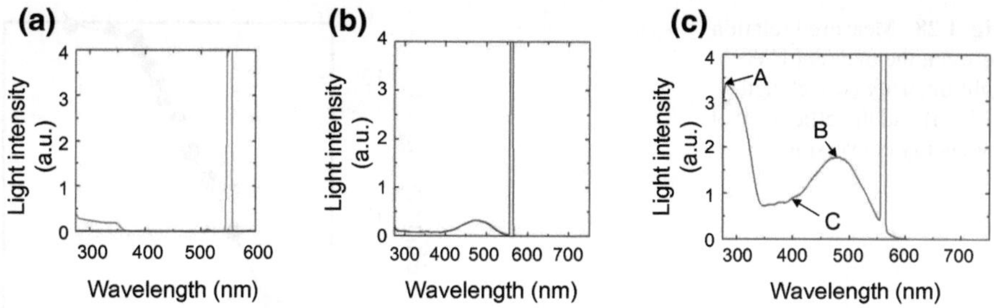


**Fig. 1.27** Excitation spectra. **a** and **b** The results acquired when only the ZnO and DCM particles are dispersed in the film, respectively. **c** The result when both ZnO and DCM particles are dispersed

to visible light (wavelength 560 nm) energy in Fig. 1.27c is estimated to be at least 10-times higher than that in Fig. 1.27a or b.

Furthermore, a bump A on the curve in Fig. 1.27c represents that the efficiency of the visible light emission is selectively enhanced when the incident UV light is resonant with the energy level $|a_{ZnO}\rangle$ of the ZnO NP. Furthermore, the bumps B and C also represent that the efficiency is selectively enhanced when the incident light is resonant with the energy level $|b_{ZnO}\rangle$ or $|c_{ZnO}\rangle$ of the ZnO NP. These selective enhances are due to the DP energy transfer to the energy levels $|b_{DCM}\rangle$ or $|d_{DCM}\rangle$ of the DCM NPs, and subsequent relaxation to the energy level $|e_{DCM}\rangle$.

Figure 1.28 shows the measured relation between the incident UV light intensity (wavelength 325 nm) and the efficiency of the energy conversion to visible light (wavelength 560 nm). It shows that the measured efficiency is proportional to the square of the incident UV light intensity.[1] This is because the conversion efficiency is proportional to the product of the numbers of DPs created in the ZnO and DCM NPs.

Figure 1.29 shows the temporal evolution of the light intensity emitted from the film when a propagating light pulse (pulse width: 2 ps) was applied to the NPs. Figure 1.29a, b were acquired when only the ZnO and DCM NPs were respectively dispersed in the film. They are nothing more than the intensities of the light emitted from the ZnO and DCM NPs, respectively, as a result of conventional fluorescence.

The rise time $\tau_r$ of the light intensity is as short as 10 ps in Fig. 1.29a, b. On the other hand, the fall time $\tau_f$ is as long as 15 and 1.4 ns, respectively, which corresponds to the radiative relaxation rate.

Figure 1.29c shows the results acquired when both ZnO and DCM NPs are dispersed in the film. Figure 1.29d is a schematic explanation of the curves in Fig. 1.29c. The time constant $\tau_r$ in this figure represents the rise time. Two constants $\tau_{f1}$ and $\tau_{f2}$ represent two different fall times. Figure 1.29c shows that the values of $\tau_r$ and

[1] Figure 1.28 shows that the efficiency saturates when the UV light intensity is higher than 0.2 W/m$^2$. Since the UV light component in sunlight is higher than 0.2 W/cm$^2$, this saturation is advantageous for the solar cell battery application to be reviewed in Sect. 1.7.3 because the conversion efficiency is maintained constant even if the incident sunlight intensity may vary from hour to hour during daytime.

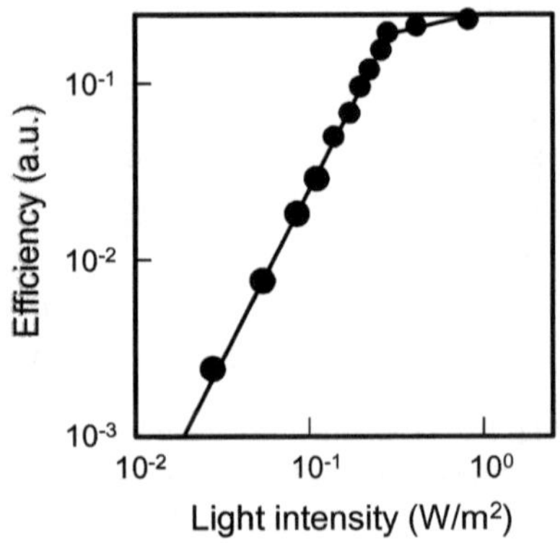


**Fig. 1.28** Measured relation between the incident UV light intensity (wavelength 325 nm) and the efficiency of the energy conversion

$\tau_{f1}$ are 100–150 ps. These correspond to the DP energy transfer times from the ZnO to DCM NPs, which are much longer than the fluorescence rise times in Fig. 1.29a, b.

Similar to the fall time of the curves in Fig. 1.29a, b, the fall time $\tau_{f2}$ is 15 ns, which is much longer than $\tau_{f1}$. This fall time $\tau_{f2}$ corresponds to the value of the radiative relaxation rate. It means that, after the DP energy transfer, conventional florescence occurs in the case where both the ZnO and DCM NPs are dispersed in the film. It should be noted that Fig. 1.29a, b do not have the temporal behaviors represented by the time constants $\tau_r$ and $\tau_{f1}$.

The temporal evolutions shown in Figs. 1.24 and 1.29c have several common features even though the materials used are different:

(1) The temporal evolution expressed as $\exp(-t/\tau_{f1})$ and the nutation behavior in Fig. 1.24 have the same origin as that of the time evolution represented by $\tau_{f1}$ in Fig. 1.29c, namely, the DP energy transfer between NPs. The rise time also originates from this transfer.
(2) The temporal evolution expressed as $\exp(-\sqrt{t/\tau_{f2}})$ in Fig. 1.24 has the same origin as that of the time evolution represented by $\tau_{f2}$ in Fig. 1.29c, namely, the radiative relaxation in each NP.

Features (1) and (2) above represent unique phenomena which are different from each other. The former is exactly the novel off-shell scientific phenomenon (Fig. 1.15). The latter is no more than a conventional on-shell phenomenon. The fact that these temporal evolutions are respectively expressed as $\exp(-t/\tau_{f1})$ and $\exp(-\sqrt{t/\tau_{f2}})$ suggests that they correspond to the quantum walk [81] and the relaxation processes, respectively. As for the former process, it should not be considered as a mere random walk because its energy transportation is linearly dependent on time, not on the square root of time. It is expected that the unique autonomous spatial evolution feature described in Sect. 1.6.3 can be also analyzed in terms of a quantum walk, one fundamental feature of which is a linear dependence on time.

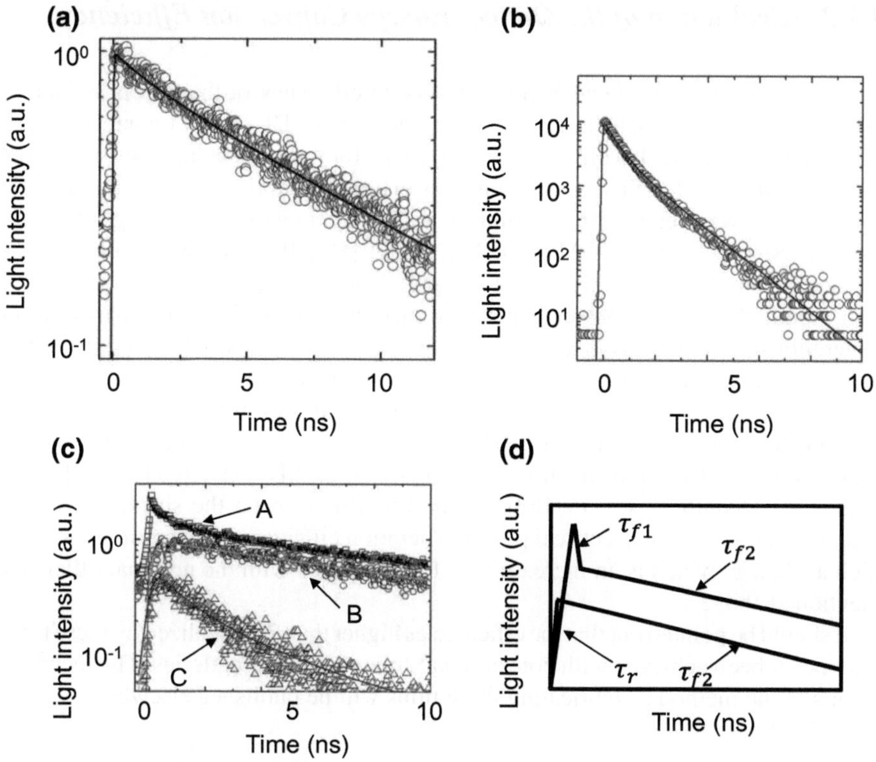


**Fig. 1.29** Temporal evolutions of the light intensity emitted from the film. **a** The result acquired when only the ZnO NPs are dispersed in the film. The vertical axis represents the light intensity (wavelength 350 nm) emitted from the electron in the lower energy level $|a'_{\mathrm{ZnO}}\rangle$ of the ZnO by illuminating the film with light (wavelength 340 nm), which is resonant with the higher energy level $|a_{\mathrm{ZnO}}\rangle$. **b** The result acquired when only the DCM NPs are dispersed in the film. The vertical axis represents the light intensity (wavelength 560 nm) emitted from the electron in the energy level $|e_{DCM}\rangle$ of the DCM by illuminating the film with light (wavelength 340 nm). **c** The result acquired when both the ZnO and DCM NPs are dispersed in the film. Curve A represents the light intensity (wavelength 350 nm) emitted from the electron in the lower energy level $|a'_{\mathrm{ZnO}}\rangle$ of the ZnO by illuminating the film with light (wavelength 340 nm), which is resonant with the energy level $|a_{\mathrm{ZnO}}\rangle$. Curve B represents the light intensity (wavelength 560 nm) emitted from the electron in the energy level $|e_{DCM}\rangle$ of the DCM by illuminating the film with light (wavelength 340 nm), which is resonant with the energy level $|a_{\mathrm{ZnO}}\rangle$ in the ZnO. Curve C represents the light intensity (wavelength 420 nm) emitted from the electron in the energy levels $|b_{\mathrm{ZnO}}\rangle$ and $|d_{DCM}\rangle$ of the ZnO and DCM, respectively, by illuminating the film with light (wavelength 420 nm), which is resonant with the energy level $|c_{\mathrm{ZnO}}\rangle$ in the ZnO. **d** Schematic explanation of the curves in **c**

### 1.7.3 Evaluation of the Optical Energy Conversion Efficiency

It is expected that the optical energy conversion efficiency of the film is as high as 90–95% by referring to the large difference between the DP energy transfer time ($\tau_r$, $\tau_{f1} = 100$–$150$ ps) and the lifetime ($\tau_{f2} = 15$ ns) for fluorescent light emission from the DCM NP. Based on this expectation, a novel film was fabricated for converting UV light energy (wavelength 300–350 nm) to visible light energy (wavelength 560 nm) by dispersing ZnO and DCM NPs (Fig. 1.30a) [82]. By putting this film on the surface of a commercially available Si solar cell battery (surface area 156 mm × 156 mm, nominal electrical power generation efficiency 18.1%), the electrical power generation efficiency was measured to be as high as 20.0%, which is an increase of 1.9% compared with the nominal efficiency mentioned above.

To realize an even higher increase, another type of film was recently developed by replacing the DCM NPs with BBQ dye NPs (Fig. 1.30b), which allowed UV light in the wavelength range 300–350 nm to be efficiently converted to visible light with a wavelength of 450 nm [83]. By putting this film on the surface of the Si solar battery above, the electrical power generation efficiency was measured to be as high as 20.2%, which is an increase of 2.1% compared with the nominal efficiency mentioned above.

It should be pointed out that the efficiencies higher than 20% realized by these films have never been achieved with conventional solitary Si solar batteries. The technical details of the method of fabricating these films will be published elsewhere.

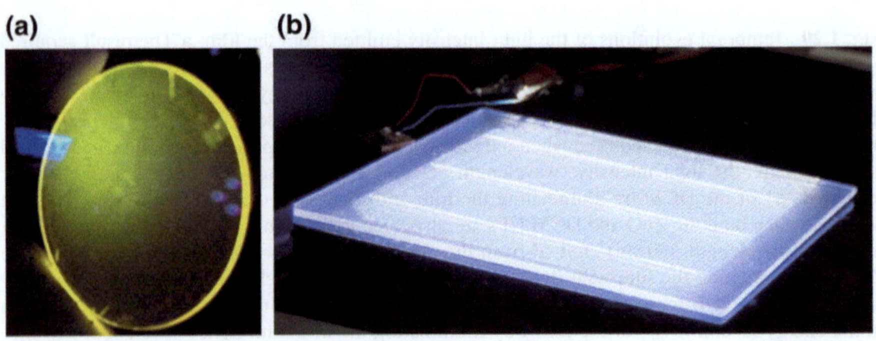

**Fig. 1.30** Photographic images of the film. The film is 10 μm thick and is coated on the front surface of a glass substrate. After an index-matching oil is coated on the rear surface, it is placed on a solar cell battery surface. **a** ZnO and DCM NPs are dispersed in the film. **b** ZnO and BBQ NPs are dispersed in the film

## 1.8 Toward Step 3

This section presents reasons why step 3 in Sect. 1.4 is required. To proceed with step 3, a novel theory that describes the micro-macro duality of quantum fields is essential. In order to develop such a theory, this section and Sect. 1.9 review the requirements in theoretical studies, which have been identified from experimental results accumulated for more than a quarter of a century. For this review, a fiber probe (Fig. 1.1a) is adopted as a representative device to create and measure the DP [6]. Since this device has a simple structure, it should be possible to analyze the transformation of an electromagnetic (EM) field between micro- and macroscopic systems, which will open up a new field of off-shell science.

This fiber probe has been used at the heart of a novel microscope and spectrometer that exhibit ultrahigh spatial resolution beyond the diffraction limit of propagating light [84]. Conformations and structures of a variety of nanometric materials have been measured and analyzed with these instruments, which have been commercially available [85].

### *1.8.1 Principles of Creation and Measurement of Dressed Photons*

In order to identify requirements from an experimental viewpoint, this subsection reviews the principles of creation and measurement of DPs. The structures and performance levels of fiber probes are also reviewed. Also, it is pointed out that a fiber probe can be replaced by a NP. Finally, requirements for developing novel theories are presented.

As schematically illustrated in Fig. 1.15, the DP is a quantum field that exists in an off-shell area that is displaced from the shell of the dispersion relation between energy and momentum. Since the uncertainty $\Delta p$ of the momentum $p$ is large in this area ($\Delta p \gg p$), the size $a$ of the DP is much smaller than the wavelength $\lambda$ of the propagating light ($a \ll \lambda$) due to the Heisenberg uncertainty principle. In order to create such a small DP, a fiber probe has been used, as is schematically explained by Fig. 1.31a. The operating mode of the fiber probe in this figure is called the illumination mode (I-mode) [86], in which the tail of the fiber probe is illuminated with propagating light (the EM field on shell) to create a DP at the nanometric tip of the probe.

Since the created DP is localized on the tip, it is measured by inserting a sonde into the DP for acquiring the response from the DP. That is, the DP is measured by acquiring its response to a stimulus applied from the outer system. A NP has been used as such a sonde (Fig. 1.31b). By putting this NP in close proximity to the tip of the fiber probe, the DP energy is exchanged between the fiber probe tip and the NP, resulting in excitation of an electron in the NP. The excited electron can create a photon. Since this photon is a conventional scattered light field on shell, it can be

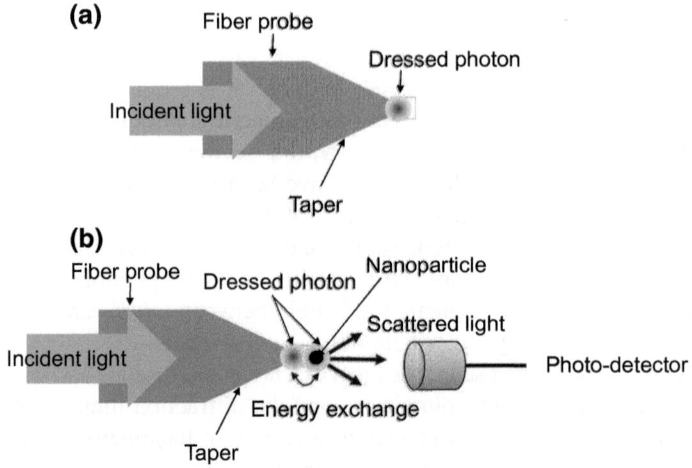


**Fig. 1.31** Illumination mode for the fiber probe operation. **a** For creating the DP. **b** For measuring the DP

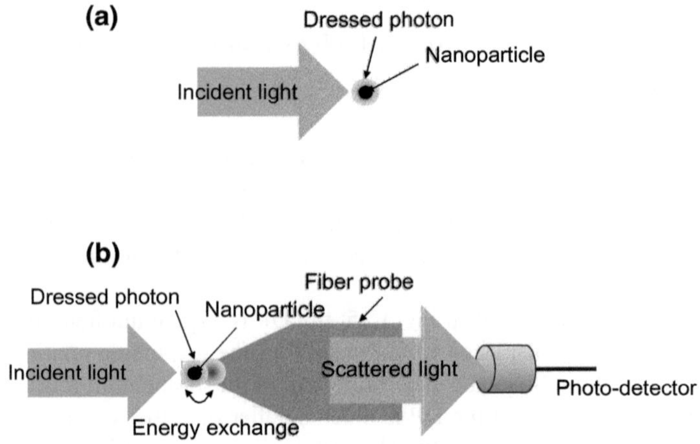


**Fig. 1.32** Collection mode of the fiber probe operation. **a** For creating the DP. **b** For measuring the DP

measured by a conventional photo-detecting device, and thus, the response can be acquired. In this I-mode, the fiber probe and NP respectively play the roles of a light source and a detector for creating and measuring the DP.

It is possible to exchange the roles of the fiber probe and the NP: First, the NP is illuminated by propagating light to create the DP (Fig. 1.32a). Next, a fiber probe, which is used as a sonde, is brought close to the NP (Fig. 1.32b). The DP energy is thus exchanged between the NP and fiber probe tip, resulting in excitation of an electron in the tip of the fiber probe. The excited electron can create a photon, i.e., scattered light. Since this scattered light is guided through the fiber probe and reaches its tail, it can be measured by a conventional photo-detecting device, and thus, the response can be acquired. The operating mode of the fiber probe in this figure is named the collection mode (C-mode) [86].

### *1.8.2 Performance of Fiber Probes*

As summarized in Fig. 1.1a, a variety of fiber probes have been developed so far, some of which have been commercially available [6]. The size and conformation of the tip and taper of the fiber probe have been empirically controlled during the fabrication process, resulting in sufficiently high efficiencies for creating and measuring DPs for practical use. These high efficiencies are indispensable for reliable conversion of the EM field from macro- to microscopic systems and also from micro- to macroscopic systems, respectively, in the case of the I- and C-modes.

Figure 1.1b represents the efficiency of measuring the DP acquired when the fiber probe was used in the I-mode [87]. This efficiency is expressed as the "throughput", which is the ratio between the measured optical power and the optical power incident at the tail of the fiber probe. The horizontal axis is the diameter $d_f$ of the foot of the taper protruding from an opaque metallic film, which was deposited for blocking unwanted scattered light. In these old experimental results acquired more than 15 years ago, it should be pointed out that a certain amount of unwanted scattered photons was measured simultaneously with the DP when $d_f$ was larger than 100 nm. This signal mixing was due to insufficient shielding resulting from the immature metallic film coating technology available at the time.

It should be noted that the spatial resolution of this novel microscope and spectrometer are determined by the value of the tip radius, $a_p$. DPs have been used in these instruments with ultrahigh spatial resolution beyond the diffraction limit of light. The NPs in Figs. 1.31a and 1.32a correspond to the specimens to be measured by these instruments. In the I-mode, the DP on the tip of the fiber probe serves as a light source to illuminate the specimen. In the C-mode, the light source is the DP on the NP, which is picked up by the fiber probe.

In order to analyze the spatial resolution and contrasts of the acquired microscopic and spectroscopic images, the cross-sectional profile of the fiber probe is shown in Fig. 1.33a, as was given in Fig. 1.1b [88]. Here, $a_p$ is the tip radius, $d_f$ is the diameter of the foot of the taper protruding from an opaque metallic film, which is deposited for blocking the propagating scattered light, and $\theta$ is the cone angle. The spatial resolution of the measurement is governed by the size of the DP created on the tip, which is equivalent to the tip radius $a_p$, because the spatial profile of the DP field is represented by the Yukawa function (1.8).

Here, the contributions from the DPs created on the taper have to be also considered for evaluating the contrast of the acquired image. For this consideration, Fig. 1.33b schematically explains that the taper is approximated as a chain of small spheres which are connected in order to increase the radius, from $a_p$ up to $d_f/2$, as

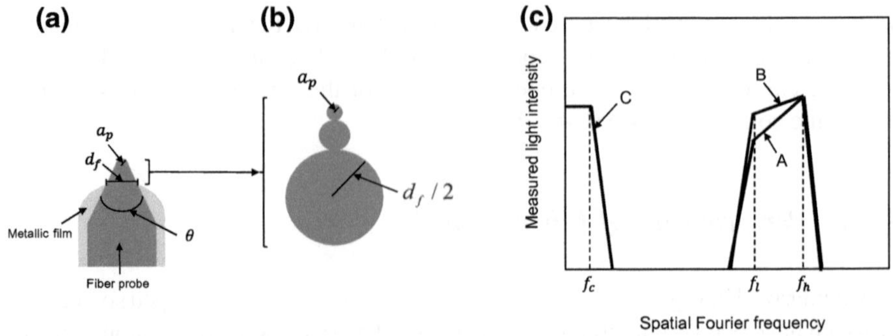


**Fig. 1.33** A fiber probe and spatial Fourier spectra of the acquired image. **a** Cross-sectional profile of the fiber probe. **b** A taper approximated by a chain of spheres. **c** Spatial Fourier spectrum of the image acquired in the Collection mode. Lines A and B correspond to the fiber probe with small and large cone angles $\theta$, respectively. Line C represents the spectrum acquired by a conventional optical microscope

shown in Fig. 1.39. These spheres receive the DP energy created on the NP in the case of the C-mode, whereas they create the DP on their surfaces in the case of the I-mode. Therefore, in the case of the C-mode, due to the size-dependent resonance feature of the DP energy exchange [29], high-efficiency measurement is achieved if the size of the DP on the NP falls between $a_p$ and $d_f/2$. This means that this efficiency has the characteristics of a spatial band-pass filter. Its spatial Fourier spectra are shown by the lines A and B in Fig. 1.33c. They show that the C-mode can detect a DP whose size falls within the pass band of this spatial filter (i.e., between $a_p$ and $d_f/2$). Its high-frequency cutoff $f_h$ is proportional to $a_p^{-1}$, showing that the spatial resolution is determined by $a_p$. On the other hand, the low-frequency cutoff $f_l$ is proportional to $d_f^{-1}$.

Figure 1.33c shows that with a smaller $\theta$ (line A), a lower efficiency is exhibited at $f_l$ than that with a larger $\theta$ (line B), because the sphere of radius $d_f/2$ is farther from the tip in the case of a smaller $\theta$. This means that the sharper fiber probe can achieve higher selectivity in measuring the DP with a size as small as the tip radius $a_p$. In other words, the high-spatial-resolution component in the image is acquired with higher contrast. In the case of the I-mode, the lines A and B represent the size dependence of the DP energy created on the fiber probe; that is, a DP with a size ranging from $a_p$ to $d_f/2$ is efficiently created.

A conventional optical microscope collects propagating light scattered from the specimen with convex lenses for acquiring an image of the specimen. The spatial Fourier spectrum of the acquired image is represented by the line C in Fig. 1.33c. Due to the diffraction limit of light, it has the characteristics of a low-pass filter, whose high-frequency cutoff $f_c$ is determined by the wavelength $\lambda$ of the light, i.e., is proportional to $\lambda^{-1}$.

By comparing the lines A, B, and C, it is confirmed that the spatial resolution of the microscope using DPs is much higher than that of the conventional microscope because $a_p \ll \lambda$ ($f_h \gg f_c$), which is the origin of the name "ultrahigh spatial resolution microscope".

It should be noted here that the spatial Fourier spectral characteristics of the microscopic images acquired by this ultrahigh spatial resolution microscope and the conventional optical microscope do not have any strong correlation between each other. In other words, the images acquired by these microscopes are totally different from each other, in addition to the ultrahigh spatial resolution feature of the former. This is because their spatial filtering characteristics are different; i.e., band-pass filtering and low-pass filtering. By noting the size-dependent resonance feature of the DP energy exchange [10], it should be pointed out that the microscope using DPs acquires an image of the fiber probe tip itself instead of an image of the specimen NP.

### *1.8.3 Using Nanoparticles*

A novel method has been developed recently in order to replace the role of the fiber probe operating in the I-mode by an NP, as is schematically explained by Fig. 1.34a. In this scheme, an NP is illuminated by propagating light to create a DP. One may worry that the creation efficiency will be very low because the interface between the macro- and microscopic systems, i.e., the taper of the fiber probe in Fig. 1.2, is missing in this configuration. However, novel interface devices, such as an optical nano-fountain and an optical energy transmitter (Sects. 1.6.1 and 1.6.2, respectively), have been developed by using multiple NPs, enabling drastic increases in efficiency.

The replacement mentioned above has been realized also for the C-mode: As is schematically explained by Fig. 1.34b, in this scheme, multiple NPs are arranged around the NP on which the DP is originally created. As a result, the DP energy is exchanged between the center NP and the surrounding multiple NPs to excite an electron. The excited electron can subsequently create scattered light, which can be acquired by a conventional photo-detecting device.

It should be pointed out that a specific phenomenon of autonomous DP energy exchange has been observed among multiple randomly arranged NPs [14]. That is, it has been experimentally confirmed that the optimum route was autonomously selected for the DP energy transfer in the optical nano-fountain and optical energy transmitter devices.

### *1.8.4 Requirements for Novel Theories*

The gray cone in the phase diagram of Fig. 1.35 represents the area in which the DP measuring efficiency is high, which was empirically found through experimental work on fabricating and using an I-mode fiber probe. Here, $a_p$ is the tip radius, $a_s$ is the radius of the spherical NP, $a_s/a_p$ is their ratio, and $\theta$ is the cone angle of the

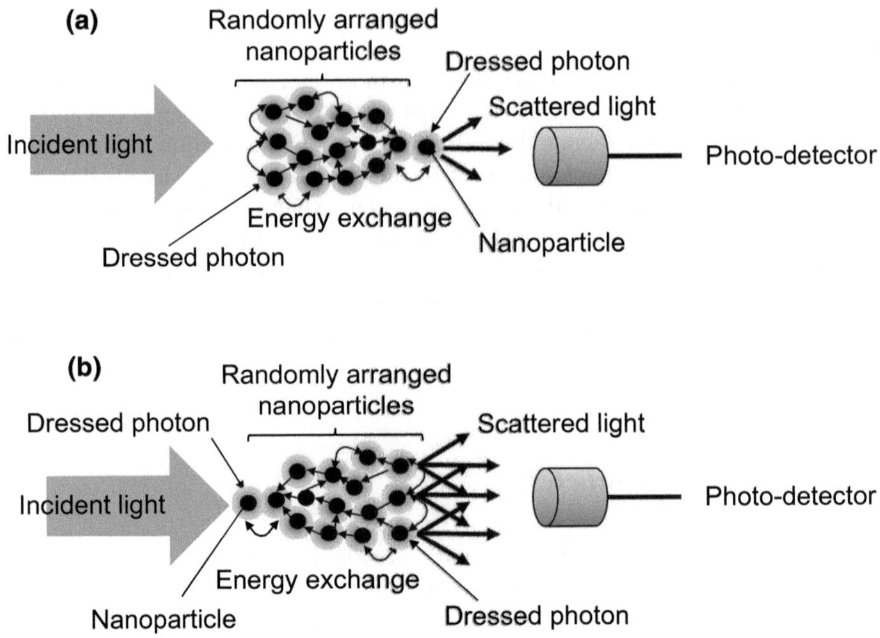


**Fig. 1.34** Nanoparticles (NPs) for creation and measurement of the DP. The multiple randomly arranged NPs correspond to the taper of the fiber probe. **a** For creating the DP. **b** For measuring the DP

taper. It should be pointed out that the efficiency is the highest when $a_s/a_p = 1$, which is due to the size-dependent resonance feature of the DP energy exchange [29]. A similar conical area can be derived also for the C-mode. A novel theory is required since one of the major requests from experimentalists is to find the optimum condition for realizing the highest efficiency of creation and measurement of DPs. It is expected that Fig. 1.35 will serve as a reference to find this condition.

To find the optimum condition, it should be also noted that the detailed profile of the tip and taper of the fiber probe are not smooth but have some roughness on their surfaces, as is schematically shown in Fig. 1.36. Specifically, Fig. 1.36a represents a conical surface profile with a smooth surface, which can be observed from a far field view point. However, the NP in Fig. 1.31b, which is installed in the near field of the fiber probe surface, may see a magnified surface and find a lot of bumps (Fig. 1.36b), on which multiple DPs with a variety of sizes are created. That is, a hierarchy exists in the DP measurement, which depends on the position and size of the NP to be used as a sonde for the measurement. A novel theory that can describe this hierarchy, as well as the autonomy mentioned in Sect. 1.8.3, is needed.

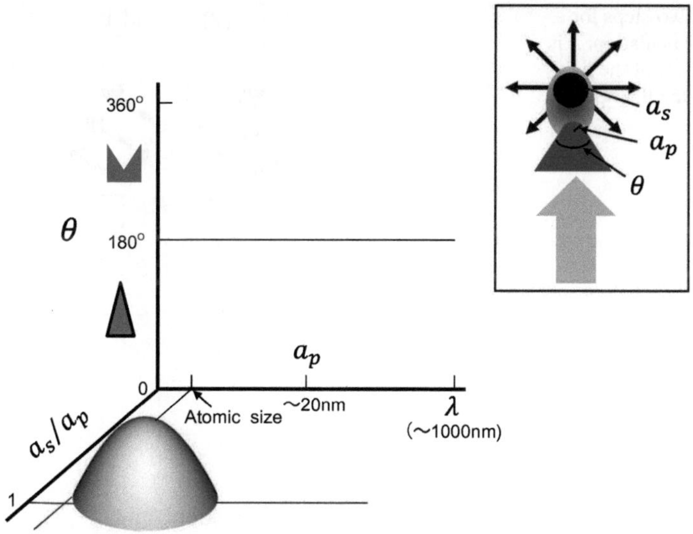


**Fig. 1.35** Phase diagram for representing the area in which the efficiency of measuring the DP is high. $\lambda$ is the wavelength of the light incident on the fiber probe

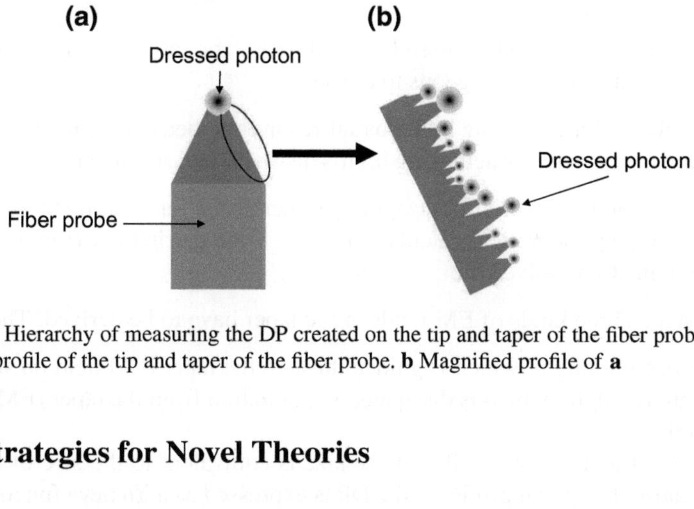


**Fig. 1.36** Hierarchy of measuring the DP created on the tip and taper of the fiber probe. **a** Cross-sectional profile of the tip and taper of the fiber probe. **b** Magnified profile of **a**

## 1.9 Strategies for Novel Theories

This section presents problems to be solved for proceeding with step 3 in Sect. 1.4, which have been identified by experimentalists. Some promising novel theoretical methods for solving these problems are reviewed.

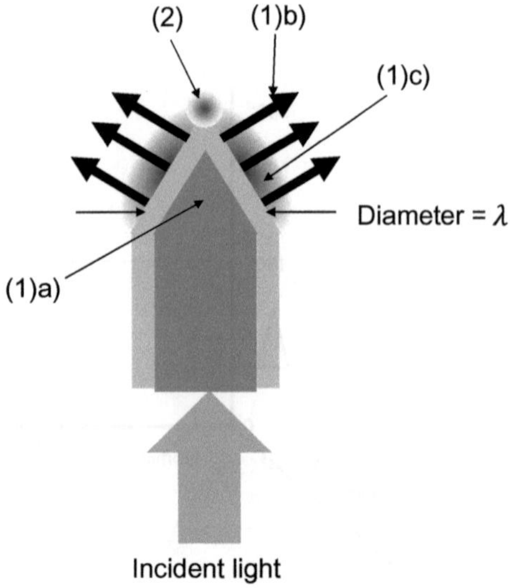


**Fig. 1.37** Two-steps for theoretical calculation. $\lambda$ is the wavelength of the light incident on the fiber probe

### *1.9.1 Problems to Be Solved*

In the case where the I-mode is used by illuminating the tail of the fiber probe with propagating light, it is advantageous to create:

(A) a small DP for achieving high-spatial-resolution measurement, and
(B) a high energy DP for achieving high-sensitivity measurement.

In order to find the criteria for designing a fiber probe for creating these DPs, the following two-step theoretical calculation ought to be carried out (Fig. 1.37). That is, the problems to be solved are

(1) 1st step: Three kinds of EM fields in the taper have to be derived. They are
  (a) Propagating light, which is guided through the taper (EM field on shell).
  (b) Scattered light, which is dissipated via radiation from the taper (EM field on shell).
  (c) A DP (EM field off shell), whose size is equivalent to the size of the taper because the spatial profile of the DP is expressed as a Yukawa function (1.8).

(2) 2nd step: The DP on the tip of the fiber probe has to be derived.

The DP in (2) is created as a result of the DP energy transfer from the fields (1a) and (1c). It means that the conformation and structure of the taper play essential roles in creating the DP of (A) and (B) on the tip. In other words, it is essential to optimize the magnitude of the energy dissipated from the taper, which is the magnitude of the energy of the field (1b).

It should be pointed out that this taper is the interface between micro- and macroscopic systems (i.e., between the tip and the tail of the fiber probe), and thus, it plays an essential role in the micro-macro duality However, EM fields in the taper have never been correctly described by the conventional EM field theories because these theories cannot be applied to the taper due to its sub-wavelength diameter. Namely, a method of numerical analysis based on Maxwell's equations (for example, the Finite Difference Time Domain (FDTD) method) is not suitable for deriving the EM field of (1b) and (1c) [89]. In particular, in the case of (1c), the use of the FDTD method is useless because it does not take into account the longitudinal component of the electric field [53].

### *1.9.2 Expected Theoretical Methods*

Several novel theoretical methods have been proposed to solve the problems presented in Sect. 1.9.1. This section reviews these methods.

Figure 1.35 shows that the highest efficiency was obtained in the case of $a_s/a_p = 1$, due to the size-dependent resonance in the DP energy exchange between the fiber probe and the NP [29]. This case corresponds to the case where the magnitude of the energy dissipation from the taper (the scattered light energy of (1b) in Sect. 1.9.1) takes the minimum. Therefore, in order to find the creation methods (A) and (B) in Sect. 1.9.1, it is essential to explore the conformation and size of the fiber probe which maximize the difference between the energy of the DP localized at the tip and the energy lost due to dissipation at the taper. For this exploration, Fig. 1.38a was derived, in which the value of $a_s/a_p$ in Fig. 1.35 was replaced by the magnitude

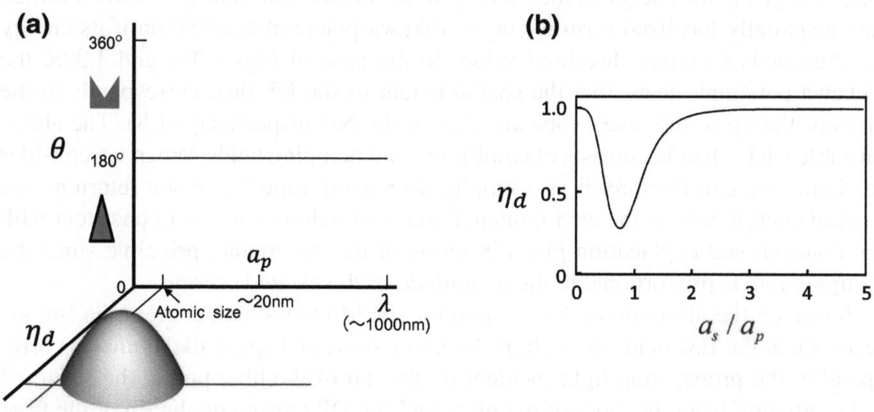


**Fig. 1.38** Phase diagram for representing the area in which the efficiency of measuring the DP is high. **a** The ratio of the sizes $a_s/a_p$ in Fig. 1.35 is replaced by the energy dissipation rate $\eta_d$. **b** The relation between $a_s/a_p$ and $\eta_d$ used for illustrating **a**

of the energy loss $E_d$ (magnitude of the light energy scattered from the taper). It should be noted that this figure uses the energy dissipation rate $\eta_d(= E_d/E_i)$, which is defined by the ratio between $E_d$ and the energy $E_i$ of the light incident on the tail of the fiber probe. For this replacement, a semi-quantitative relation between $a_s/a_p$ and $\eta_d$ was derived based on Fig. 2.6 of [29], which is shown by Fig. 1.38b.

Here, the problem is that the conventional EM field theories cannot be used to analyze the magnitude $E_d$ of energy dissipation because the taper is of sub-wavelength size. To solve this problem, it would be advantageous to use the concept of effective mass of the EM field instead of using the conventional method. This may enable estimation of the magnitude of the energy dissipation by assuming that the energy dissipation takes place during the process of transforming the massless free photon on shell to the off-shell photon with a finite mass. The Clebsch dual field theory was developed for this estimation by noting the duality between the fields in the space-like and time-like areas in the Minkowski diagram [48, 49].

The energy-momentum tensor of the Clebsch dual field is isomorphic to the left-hand side of the Einstein field equation represented as a Ricci curvature tensor defined in terms of a contracting Riemann tensor, thus suggesting an interpretation wherein the energy-momentum tensor of a virtual photon field carries a part of the energy-momentum of "vacuum", as in classical space-time. It also suggests that it may represent a so-far unidentified energy field similar to the controversial dark energy in cosmology, since, corresponding to its space-like characteristics, the associated scalar curvature is negative. According to a quantum mechanical point of view, a vacuum is considered as a fluctuating energy field where creation and annihilation of a variety of virtual particles occur incessantly.

Mathematical analysis of the Clebsch dual field shows that a space-like Clebsch field is stable as long as it exists in wave-like modes extending uniformly in space-time. However, once local perturbations arising from the interactions of material fields are given, the energy of the Clebsch field becomes unstable and is transformed into a spatially localized form having a Yukawa-potential distribution if its energy level exceeds a certain threshold value. In the case of Figs. 1.31a and 1.32a, the Yukawa-potential means that the spatial extent of the DP field corresponds to the sizes of the tip of the fiber probe and that of the NP, respectively (1.8). The above unstable Clebsch field consists of amplifying and damping fields. When we consider the latter, we can interpret it as a rapidly decreasing time-like mode returning the excited energy back to the environmental material field in a manner consistent with the conventional explanation given in terms of the uncertainty principle since the damping rate is proportional to the magnitude of the excited energy.

Based on the discussions above, applying the Minkowski diagram to the I-mode reveals that the EM fields in the light-like, time-like, and space-like domains correspond to the propagating light incident on the tail of the fiber probe, the scattered light radiating from the taper of the fiber, and the DP created on the tip of the fiber probe, respectively, as is schematically explained by Fig. 1.39a. In the case of the C-mode, this correspondence is also explained by Fig. 1.39b. Since the light-like field produces a pair consisting of time-like and space-like fields, the creation rate of the space-like field can be maximized by minimizing the creation rate of the time-like

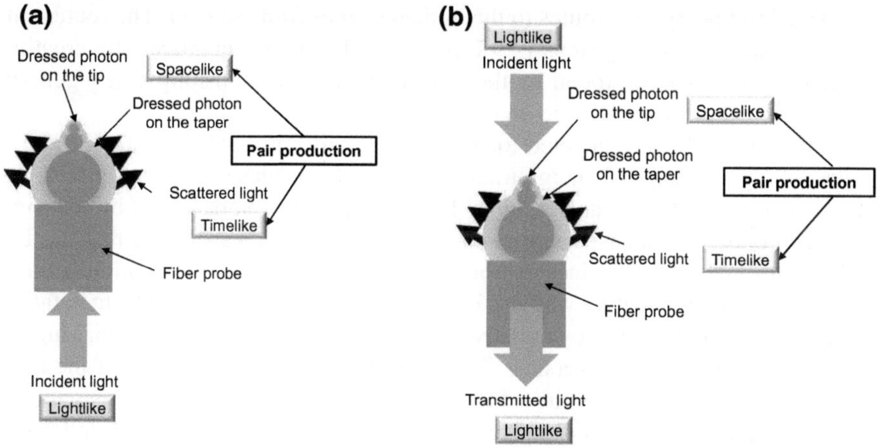


**Fig. 1.39** Electromagnetic fields in the light-like, time-like, and space-like domains. **a** Illumination mode (transformation from macro- to microscopic systems). **b** Collection mode (transformation from micro- to macroscopic systems)

field. As a result, the DPs in (A) and (B) in Sect. 1.9.1 can be efficiently created. It should be pointed out that the hierarchy explained in Fig. 1.36 can be described by including the nonlinear interaction in the theoretical approach described above.

It should be noted that the gray cones in Figs. 1.35 and 1.38a have asymmetric profiles. This is because the DP is intrinsically created on the tip of the fiber probe, which has a translationally asymmetric profile. Theoretical models for the off-shell EM field should be developed based on this asymmetry. For this advanced theoretical model, use of the quadrality scheme [51] based on category theory is expected to be a promising approach for describing the transformation of the microscopic DP to the macroscopic system. Furthermore, a novel measurement theory should be constructed by noting that the fiber probe and NP in Figs. 1.31b and 1.32b form a composite system originating from the DP energy exchange. Tomita's decomposition theorem [90] is expected to be promising for this construction. Further advances in these theoretical studies are expected to lead to developments in the field of off-shell science.

## 1.10 Summary

After the introduction, Sects. 1.2–1.4 of this article reviewed the experimental and theoretical studies of DPs carried out in the last three decades. It was pointed out that the main effort in recent years has focused on experimental studies, resulting in the development of generic technologies for supporting modern society. By referring to these drastic experimental advances, it was also pointed out that further effort should be devoted to finding new routes to theoretical studies from now on. The routes can be developed by following steps 1–3. It is expected that a novel process for creating or detecting DPs will be found by these developments, establishing a new generic science, named off-shell science, in the near future.

Sections 1.5–1.7 were devoted to proposing how to proceed with steps 1 and 2 by describing a unique DP energy transfer process between semiconductor NPs. As a first application of this transfer, novel DP devices were demonstrated by focusing on a nano-optical condenser and an energy transmitter. It was shown that the DP energy transfer in these devices exhibited an autonomous spatial evolution feature. A temporal evolution feature was also demonstrated, which originated from nutating DP energy transfer followed by radiative relaxation. As a second application, a highly efficient optical energy conversion film was demonstrated. The origin of the high conversion efficiency was confirmed to be equivalent to that of the DP devices above. It was suggested that the spatial and temporal evolutions above can be analyzed based on theoretical models involving a quantum walk and a random walk. As an application of this film, a drastic increase in electrical power generation efficiency of a silicon solar battery was demonstrated.

Sections 1.8 and 1.9 reviewed how to proceed in step 3 by adopting a fiber probe in order to identify the requirements for theoretical studies. The operating principles, structures, performance levels of fiber probes were reviewed. The spatial resolution and contrast of the image acquired by a microscope and spectrometer using such a fiber probe were also explained. It was pointed out that the fiber probe could be replaced by an NP, and that multiple NPs exhibited a specific phenomenon of autonomous energy transfer of the DP. A phase diagram was empirically derived through experiments in order to show the experimental conditions necessary for achieving high-efficiency measurements of the DP. Based on this diagram, requirements for a novel theory were identified to find the optimum conditions giving the highest efficiency. Furthermore, it was required that this theory should be able to describe the autonomy above, and also the hierarchy observed in DP measurement.

To meet these requirements, some promising novel theoretical approaches were reviewed. One was an approach based on Clebsch dual field theory. Uses of the quadrality scheme based on category theory and a novel measurement theory were also presented as promising approaches for analyzing the detailed physical properties of DPs, leading to the development of off-shell science.

**Acknowledgements** The author thanks Drs. I. Ojima, H. Sakuma (Research Origin for Dressed Photon), I. Banno (Yamanashi Univ.), H. Saigo (Nagahama Inst. Bio-Science and Technol.), K. Okamura (Nagoya Univ.), H. Ando (Chiba Univ.), S. Sangu (Ricoh Co., Ltd.), and T. Kawazoe (Tokyo Denki Univ.) for their valuable comments and discussions.

Some of the work described in this article was supported by the Research Foundation for Opto-Science and Technology.

## References


1. E.H. Synge, Phil. Mag. **6**, 356 (1928)
2. H. Bethe, Phys. Rev. **66**, 163 (1944)
3. C.J. Bouwkamp, Philips Res. Rep. **5**, 401 (1950)
4. E.A. Ash, G. Nicholls, Nature **237**, 510 (1972)
5. D.W. Pohl, D. Courjon (ed.), *Near Field Optics* (Kluwer, Dordrecht, 1993), pp. 1–324
6. M. Ohtsu (ed.), *Near-Field Nano/Atom Optics and Technology* (Springer, Tokyo, 1998), pp. 31–100
7. M. Ohtsu (ed.), *Near-Field Nano/Atom Optics and Technology* (Springer, Tokyo, 1998), pp. 101–152
8. For example, *JASCO Catalogue for Scanning Near-Field Optical Microspectrometer (NFS Series)* (JASCO, Tokyo, 2003)
9. Optical Society of Japan (organizer), in *The 14th International Conference on Near-Field Optics, Nanophotonics, and Related Techniques*, Shizuoka, Japan, 4–8 September 2016
10. Condensed Matter Research Institute, Seoul National University (organizer), in *The First Asia-Pacific Workshop on Near Field Optics*, Seoul, Korea, 17–18 August 1996
11. National Chung-Kung University (organizer), in *The 11th Asia-Pacific Conference on Near-Field Optics*, Tainan, Taiwan, 10–13 July 2017
12. M. Ohtsu, *Dressed Photons* (Springer, Heidelberg, 2014), pp. 1–324
13. M. Ohtsu, Off-shell archive, OffShell: 1709R.001.v1 (2017)
14. M. Ohtsu, T. Kawazoe, H. Saigoh, Off-shell archive, OffShell: 1710R.001.v1 (2017)
15. M. Ohtsu, H. Sakuma, Off-shell archive, OffShell: 1712R.001.v1 (2017)
16. M. Ohtsu, *Dressed Photons* (Springer, Heidelberg, 2014), pp. 89–214
17. T. Kawazoe, T. Takahashi, M. Ohtsu, Appl. Phys. B **98**, 5 (2010)
18. T. Matsumoto, K. Nakamura, T. Nishida, H. Hieda, A. Kikitsu, K. Naito, T. Koda, Thermally assisted magnetic recording on a bit-patterned magnetic medium using a near-field optical head with a beaked metallic plate, in *International Symposium on Optical Memory 2007 (ISOM'07)*, Singapore, 21–25 October 2007, pp. 90–91
19. T. Yatsui, W. Nomura, M. Naruse, M. Ohtsu, J. Phys. D **45**, 475302 (2012)
20. M. Ohtsu, Silicon light emitting diodes and lasers using dressed photons, in *Progress in Nanophotonics 3*, ed. by M. Ohtsu (Springer, Heidelberg, 2015), pp. 1–56
21. M. Ohtsu, *Dressed Photons* (Springer, Heidelberg, 2014), pp. 89–136
22. M. Naruse, T. Miyazaki, T. Kawazoe, K. Kobayashi, S. Sangu, F. Kubota, M. Ohtsu, IEICE Trans. Electron. **E88–C**, 1817 (2005)
23. H. Fujiwara, T. Kawazoe, M. Ohtsu, Appl. Phys. B **100**, 85 (2010)
24. M. Naruse, M. Aono, S.-J. Kim, T. Kawazoe, W. Nomura, H. Hori, M. Hara, M. Ohtsu, Phys. Rev. B **86**, 125407 (2012)
25. T. Kawazoe, H. Fujiwara, K. Kobayashi, M. Ohtsu, IEEE J. Select. Top. Quantum Electron. **15**, 1380 (2009)
26. N. Tate, Y. Liu, T. Kawazoe, M. Naruse, T. Yatsui, M. Ohtsu, Appl. Phys. B **110**, 39 (2013)
27. M. Ohtsu, *Dressed Photons* (Springer, Heidelberg, 2014), p. 3
28. M. Ohtsu, *Dressed Photons* (Springer, Heidelberg, 2014), p. 31
29. M. Ohtsu, *Dressed Photons* (Springer, Heidelberg, 2014), pp. 33–36
30. K. Kobayashi, Y. Tanaka, T. Kawazoe, M. Ohtsu, Localized photon model including phonons' degrees of freedom, in *Progress in Nano-Electro-Optics VI*, ed. by M. Ohtsu (Springer, Heidelberg, 2008), pp. 41–66
31. K. Kobayashi, S. Sangu, H. Ito, M. Ohtsu, Phys. Rev. A **63**, 013806 (2001)
32. T. Kawazoe, M.A. Mueed, M. Ohtsu, Appl. Phys. B **104**, 747 (2011)
33. B. Thubthimthong, T. Kawazoe, M. Ohtsu, Spectral analysis of a high-power infrared silicon light emitting diode of dressed photons, in *Proceedings Laser Congress 2017*, Nagoya, Japan, 1–5 October 2017
34. M.A. Tran, T. Kawazoe, M. Ohtsu, Appl. Phys. A **115**, 105 (2014)

35. M. Yamaguchi, T. Kawazoe, T. Yatsui, M. Ohtsu, Appl. Phys. A **121**, 1389 (2015)
36. T. Kawazoe, M. Ohtsu, Appl. Phys. A **115**, 127 (2014)
37. K. Suzuki, K. Chieda, T. Kawazoe, T. Yatsui, M. Ohtsu, Control of emission spectrum from a green SiC-LED using dressed-photon-phonon annealing, in *Abstract of the 61st JSAP Spring Meeting*, Sagamihara, Japan, 17–20 March 2014, paper number 18a-F12-11
38. Q.H. Vo, T. Kawazoe, M. Ohtsu, Fabrication of SiC-LED using dressed-photon-phonon annealing with two light sources, in *Abstract of the 61st JSAP Spring Meeting*, Sagamihara, Japan, 17–20 March 2014, paper number 18a-F12-10
39. T. Kawazoe, K. Nishioka, M. Ohtsu, SiC light emitting diode and its polarization control using dressed photons, in *Abstract of the International Display Workshop*, Niigata, Japan, 3–5 December 2014, pp. 1061–1063
40. T. Kawazoe, K. Nishioka, M. Ohtsu, Appl. Phys. A **121**, 1409 (2015)
41. T. Kawazoe, M. Ohtsu, K. Akahane, N. Yamamoto, Appl. Phys. B **107**, 659 (2012)
42. M. Ohtsu, Y. Teramachi, T. Miyazaki, IEEE J. Quantum Electron. **24**, 716 (1988)
43. M. Ohtsu, Y. Teramachi, IEEE J. Quantum Electron. **25**, 31 (1989)
44. T. Kawazoe, K. Hashimoto, S. Sugiura, High-power current-injection type silicon laser using nanophotonics, in *Abstract of the EMN Nanocrystals Meeting*, Xi'an, China, 17–21 October 2016, pp. 9–11
45. M. Ohtsu, *Coherent Quanatum Optics and Technology* (KTK Scientific and Kluwer Academic, Tokyo and Dordrecht, 1992), pp. 49–81
46. M. Ohtsu, *Highly Coherent Semiconductor Lasers* (Artech House, Boston, 1992), pp. 7–60
47. W.P. Dumke, Phys. Rev. **127**, 1559 (1962)
48. H. Sakuma, I. Ojima, M. Ohtsu, Prog. Quantum Electron. **55**, 74 (2017)
49. H. Sakuma, I. Ojima, M. Ohtsu, Appl. Phys. A **123**, 750 (2017)
50. M. Katori, H. Kobayashi, Nonequilibrium statistical mechanical models for photon breeding processes assisted by dressed-photon-phonons, in *Progress in Nanophotonics 4*, ed. by M. Ohtsu, T. Yatsui (Springer, Heidelberg, 2017), pp. 19–55
51. I. Ojima, Micro-macro duality in quantum physics, in *Stochastic Analysis: Classical and Quantum Perspective of White Noise Theory*, ed. by T. Hida (World Scientific, Singapore, 2005), pp. 143–161
52. M. Ohtsu, I. Ojima, H. Saigo, Who has seen a free photon?—From mathematical physics to light-matter fusion technologies, in *Abstracts of the 11th Asia-Pacific Conference on Near-Field Optics*, Tainan, Taiwan, 10–13 July 2017, p. 54
53. I. Banno, M. Ohtsu, Irrationality of the permittivity in non-resonant near-field optics, in *Abstracts of the 11th Asia-Pacific Conference on Near-Field Optics*, Tainan, Taiwan, 10–13 July 2017, p. 35
54. I. Banno, M. Ohtsu: 'The nonlinear response theory for near-field optics and application to non-resonant effect, in *Abstracts of the 11th Asia-Pacific Conference on Near-Field Optics*, Tainan, Taiwan, 10–13 July 2017, p. 62
55. M. Ohtsu, *Dressed Photons* (Springer, Heidelberg, 2014), pp. 137–214
56. M. Ohtsu, T. Kawazoe, K. Kobayashi, M. Ohtsu, Appl. Phys. Lett. **86**, 103102 (2005)
57. M. Aono, S.-J. Kim, M. Naruse, M. Wakabayashi, H. Hori, M. Ohtsu, M. Hara, Langmuir **29**, 7557 (2013)
58. M. Naruse, W. Nomura, M. Aono, M. Ohtsu, Y. Sonnenfraud, A. Drezet, S. Huant, S.-J. Kim, J. Appl. Phys. **116**, 154303 (2014)
59. M. Ohtsu, *Dressed Photons* (Springer, Heidelberg, 2014), pp. 160–169
60. R.W. Hilditch, *An Introduction to Close Binary Stars* (Cambridge University Press, Cambridge, 2001)
61. T. Kawazoe, M. Ohtsu, S. Aso, Y. Sawado, Y. Hosoda, K. Yoshizawa, K. Akahane, N. Yamamoto, M. Naruse, Appl. Phys. B **103**, 537 (2011)
62. T. Kawazoe, K. Kobayashi, M. Ohtsu, Appl. Phys. Lett. **86**, 103102 (2005)
63. M. Naruse, T. Miyazaki, F. Kubota, T. Kawazoe, K. Kobayashi, S. Sangu, M. Ohtsu, Opt. Lett. **30**, 201 (2005)
64. W. Nomura, T. Yatsui, T. Kawazoe, M. Ohtsu, J. Nanophotonics **1**, 011591 (2007)

65. T. Yatsui, S. Sangu, K. Kobayashi, T. Kawazoe, M. Ohtsu, J. Yoo, G.-C. Yi, Appl. Phys. Lett. **94**, 083113 (2009)
66. M. Naruse, H. Hori, K. Kobayashi, T. Kawazoe, M. Ohtsu, Appl. Phys. B **102**, 717 (2011)
67. H. Imahori, J. Phys. Chem. B **108**, 6130 (2004)
68. M. Naruse, T. Kawazoe, R. Ohta, W. Nomura, M. Ohtsu, Phys. Rev. B **80**, 125325 (2009)
69. N. Johnson, *Simply Complexity* (Oneworld Publications, Oxford, 2007)
70. W. Nomura, T. Yatsui, T. Kawazoe, M. Naruse, M. Ohtsu, Appl. Phys. B **100**, 181 (2010)
71. T. Kawazoe, S. Tanaka, M. Ohtsu, J. Nanophotonics **2**, 029502 (2008)
72. L.B. Kish, I.E.E. Proc, Circuits Devices Systems **151**, 190 (2004)
73. M. Naruse, P. Holmstr'om, T. Kawazoe, K. Akahane, N. Yamamoto, L. Thylen, M. Ohtsu, Appl. Phys. Lett. **100**, 241102 (2012)
74. M. Naruse, H. Hori, K. Kobayashi, M. Ohtsu, Opt. Lett. **32**, 1761 (2007)
75. M. Naruse, F. Peper, K. Akahane, N. Yamamoto, T. Kawazoe, N. Tate, M. Ohtsu, ACM J. Emerg. Technol. Comput. Syst. **8**, 1–4 (2012)
76. M. Naruse, K. Leibnitz, F. Peper, N. Tate, W. Nomura, T. Kawazoe, M. Murata, M. Ohtsu, Nano Commun. Netw. **2**, 189 (2011)
77. A. Olaya-Castro, C.F. Lee, F.F. Olsen, N.F. Johnson, Phys. Rev. B **78**, 085115 (2008)
78. S. Sangu, K. Kobayashi, T. Kawazoe, A. Shojiguchi, M. Ohtsu, Trans. Mater. Res. Soc. Jpn. **28**, 1035 (2003)
79. T. Kawazoe, A. Mizushima, K. Matsue, M. Ohtsu, A wavelength conversion film using energy transfer via dressed photon, in *Abstract of the 60th JSAP Spring Meeting*, Atsugi, Japan, 27–30 March 2013, paper number 28p-A1-11
80. T. Kawazoe, K. Matsue, M. Ohtsu, Fabrication of ZnO-QDs for wavelength conversion film using a dressed photon, in *Abstract of the 74th JSAP Autumn Meeting*, Kyoto, Japan, 16–20 September 2013, paper number 18p-C14-16
81. N. Konno, Quantum walk, in *Quantum Potential Theory*, ed. by U. Franz, M. Schürmann (Springer, Heidelberg, 2008), pp. 309–452
82. T. Kawazoe, K. Matsue, M. Ohtsu, Size control of ZnO quantum dots in a wavelength conversion film using a dressed photon for a solar cel, in *Abstract of the 61st JSAP Spring Meeting*, Sagamihara, Japan, 17–20 March 2014, paper number 18a-F12-6
83. T. Kawazoe, C. Amagai, M. Ohtsu, High-effectiveness of crystalline silicon solar cell by a wavelength conversion film using a dressed photon, in *Abstract of the 62th JSAP Spring Meeting*, Hiratsuka, Japan, 11–14 March 2015, paper number 11p-A12-7
84. Y. Narita, T. Tadokoro, T. Ikeda, T. Saiki, S. Mononobe, M. Ohtsu, Appl. Spectrosc. **52**, 1141 (1998)
85. M. Ohtsu (ed.), *Near-Field Nano/Atom Optics and Technology* (Springer, Tokyo, 1998), pp. 133–192
86. M. Ohtsu, K. Kobayashi, *Optical Near Fields* (Springer, Berlin, 2004), p. 23
87. T. Yatsui, M. Ohtsu, High-throughput probes for near-field optics and their applications, in *Progress in Nano-Electro-Optics I*, ed. by M. Ohtsu (Springer, Berlin, 2003), p. 12
88. M. Ohtsu, K. Kobayashi, *Optical Near Fields* (Springer, Berlin, 2004), pp. 29–30
89. A. Taflove, *Computational Electrodynamics (The Finite-Difference Time-Domain Method)* (Artech House, Boston, 1995)
90. O. Bratteli, D.W. Robinson, *Operator Algebras and Quantum Statistical Mechanics*, vol. 1, 2, 2nd edn. (Springer, Berlin, 1987, 1997)

# Chapter 2
# Virtual Photon Model by Spatio-Temporal Vortex Dynamics

**Hirofumi Sakuma**

**Abstract** The issue on theoretical foundation of optical near field or dressed photon (DP) as its quantum mechanical representation in the field of nanophotonics remains unsettled. Experimental data accumulated so far seem to suggest that DP possesses characteristics of off-shell virtual photons. On the basis of this observation and critical reviews of classical as well as quantum theory of light, we are now developing a novel theory describing DP. A unique feature of the theory explained here is that it takes an important role played by spacelike momentum to describe field interactions into consideration and hence it provides an extended picture of conventional virtual photon dynamics on spacetime. We show here that the new model is closely related to so-called "scalar photon" and longitudinal mode of free electromagnetic waves that are eliminated as unphysical modes in a covariant quantization of electromagnetic wave field. So, as an important preliminary step toward constructing a new model of virtual photons, we first discuss that the above prevailing claim made by a covariant quantization theory only reflects dynamical requirement on the micro quantum world and overlooks the emergence of condensed classical modes which can be understood in the framework of Ojima's Micro-Macro Duality Scheme (MMDS) providing a mathematically derived quantum-classical correspondence. As a matter of fact, the existence of those eliminated "unphysical" modes are clearly reported in the field of classical electromagnetic theory in addition to the fact that those modes are directly related to Coulomb modes playing a significant role in electromagnetic interactions. As a central theme of this article, a newly introduced Clebsch dual field of electromagnetic wave field as an extension of conventional electromagnetic one is shown to be a promising new model of DP as virtual photons. We also point out an intriguing property that extended virtual photons with spacelike property carry negative scalar curvature similar to enigmatic dark energy discussed in cosmology.

H. Sakuma (✉)
Research Origin for Dressed Photon, c/o Yokohama Technology center,
NICHIA Corporation, 3-13-19 Moriya-cho, Kanagawa-ku, Yokohama
221-0022, Japan
e-mail: sakuma@rodrep.or.jp



T. Yatsui (ed.), *Progress in Nanophotonics 5*, Nano-Optics and Nanophotonics,
https://doi.org/10.1007/978-3-319-98267-0_2

## 2.1 Introduction

The aim of this short article is to present a new theoretical model of DP, a challenging unsettled theme in the field of nanophotonics, based on a novel perspective that DP can be understood in terms of off-shell virtual photons. The basic idea of this approach has already been presented by a couple of works of Sakuma et al. [1] (hereafter SOO1) and [2] (SOO2). But since it is a developing new idea, it is worthwhile to review the overall picture of the new model at the present stage of development together with adding new findings. The first step towards understanding DP would be to clarify how its extreme localized and stationary configuration manifests itself in a nanoscale domain. It is by no means clear how the anomalous spatial structure of DP is excited by a given forcing of incident light with a specified wavelength whose diffraction limit is much larger than the spatial scale of DP. Although the field of nanophotonics was first established by Ohtsu [3, 4] to investigate this mechanism, probably due to its difficulty, research activities directly addressing this issue remains to be minor ones, while the majority focuses on the problems that can be accessible by conventional Maxwell's theory or quantum electrodynamics (QED). In SOO1 [1], we pointed out that the key ingredient that can explain the above-mentioned characteristics of DP is spacelike components arising from the field interactions between given matter and electromagnetic ones. So, an important question is to show how it actually works in describing DP dynamics, about which we are going to provide a tentative prototype model in this article.

To understand the dynamics of off-shell photons, we first need to know the essential difference between usual "real" photons and virtual ones. The clue to understand the difference lies in the perturbative calculations of QED in which virtual photons are introduced as dynamical mediators of electromagnetic interactions between charged particles. Unlike "real" photons in the transverse modes, the polarizations of virtual photons cover not only transverse but also the longitudinal as well as "temporal" (as the fourth direction in spacetime) modes, both of which are totally neglected as "unphysical modes" in the usual quantization approach. In other words, we should say that these "unphysical modes" play crucial roles in understanding the difference between "real" and virtual photons, which clearly distinguishes our present work from a previously proposed off-shell model by Kobayashi et al. [5] in which those "unphysical modes" are not included.

Although the issue on the physical interpretation of non-transverse modes of free electromagnetic waves was resolved by Ojima [6] on the basis of Micro-Macro Duality Scheme (MMDS) [7], serious confusions are still persistent in the understanding of the mutual relations between real and virtual photons. Such being the case, we start our discussion from reviewing this issue in Sect. 2.2, since Clebsch-duality of electromagnetic field discussed in SOO1 [1] essentially depends on the existence of above-mentioned "unphysical modes". In the context of Clebsch-duality, a new interpretation of the gauge condition black is introduced, which will be given in Sect. 2.3. Section 2.4 will be allotted for the explanation of Clebsch dual field and, in

Sect. 2.5, we show, as a new central result of this paper, how DPs as virtual photons manifest themselves in our model. Concluding remarks and an extended theme will be briefly given in the final Sect. 2.6.

## 2.2 On Longitudinal Electromagnetic Modes

We start from fixing conventions and defining several useful terms. By $g_{\mu\nu}(0 \leq \mu, \nu \leq 3)$, we denote metric tensor for a four dimensional (4*d*) pseudo Riemannian manifold $\mathcal{M}$ with signature $(1, -1, -1, -1)$. A covariant vector $U_\mu$ with a lower index $\mu$ is simply called a covector. A skew symmetric second-rank tensor $X_{\mu\nu} = -X_{\nu\mu}$ is referred here to as a bivector: for instance, for $X_{\mu\nu} = F_{\mu\nu} = \partial_\mu A_\nu - \partial_\nu A_\mu$: electromagnetic field strength, we have

$$F_{\mu\nu} = \begin{pmatrix} 0 & E^1 & E^2 & E^3 \\ -E^1 & 0 & -B^3 & B^2 \\ -E^2 & B^3 & 0 & -B^1 \\ -E^3 & -B^2 & B^1 & 0 \end{pmatrix} = (\boldsymbol{E}, \boldsymbol{B}). \tag{2.1}$$

A bivector $X_{\mu\nu}$ is said to be simple if it can be written as the exterior product of two vectors $2U_{[\mu}V_{\nu]}$ (or, $X = \frac{1}{2}X_{\mu\nu}dx^\mu \wedge dx^\nu = U \wedge V$) where the square bracket denotes anti-symmetrization. The totally anti-symmetric tensor $\epsilon_{\mu\nu\rho\sigma}$ is defined by $\epsilon_{\mu\nu\rho\sigma} = \epsilon_{[\mu\nu\rho\sigma]}$, $\epsilon_{0123} = 1$ and $\epsilon_{\mu\nu\rho\sigma}\epsilon^{\mu\nu\rho\sigma} = -24$. With the aid of $\epsilon_{\mu\nu\rho\sigma}$, $\eta_{\mu\nu}$ and $\eta^{\mu\nu}$, we can define the Hodge dual $({}^*F)_{\mu\nu}$ of a bivector $F_{\mu\nu}$ by

$$({}^*F)_{\mu\nu} = -\tfrac{1}{2}\epsilon_{\mu\nu}^{\ \ \ \rho\sigma}F_{\rho\sigma}; \tag{2.2}$$

$$({}^*F)^{\mu\nu} = -\tfrac{1}{2}\epsilon^{\mu\nu}_{\ \ \ \rho\sigma}F^{\rho\sigma}; \tag{2.3}$$

$$\text{e.g.,}\ ({}^*F)_{01} = F_{32}, \quad ({}^*F)_{32} = -F_{01}. \tag{2.4}$$

In terms of the above notation, we consider the field of geodesics on a given 4*d* pseudo Riemannian manifold $\mathcal{M}$. The equation describing a geodesic assumes the form of

$$(\nabla_U U)_\mu = U^\nu \nabla_\nu U_\mu = U^\nu(\nabla_\nu U_\mu - \nabla_\mu U_\nu) + \nabla_\mu(U^\nu U_\nu/2) = 0 \tag{2.5}$$

where $\nabla_X$ denotes the covariant derivative along a vector field $X = X^\mu\partial_\mu$ associated with the Levi-Civita connection. In case that $U^\mu$ denotes velocity four vector, that is to say, it can be defined as $U^\mu = dx^\mu/ds$ where $x^\mu$ are Lagrangian coordinates of a given (fluid) particle and $ds^2 = g_{\mu\nu}dx^\mu dx^\nu$ is the infinitesimal distance measured on the trajectory of the particle, the magnitude of it is respectively normalized as

$$g_{\mu\nu}U^\mu U^\nu = U_\nu U^\nu = 0; \quad g_{\mu\nu}U^\mu U^\nu = U_\nu U^\nu = \pm 1, \tag{2.6}$$

depending on whether $U^\mu$ is lightlike or not.

Now consider Maxwell equation in the vacuum:

$$0 = \nabla_\nu F^{\nu\rho} = -g^{\rho\sigma}\nabla_\sigma(\nabla_\tau A^\tau) + [g^{\sigma\tau}\nabla_\sigma\nabla_\tau A^\rho + R^\rho_\sigma A^\sigma], \tag{2.7}$$

where $R^\rho_\sigma$ denotes Ricci curvature tensor which is neglected except for cosmological arguments. So, under the Lorentz gauge condition: $\nabla_\nu A^\nu = 0$ with $R^\rho_\sigma = 0$, (2.7) reduces to

$$g^{\sigma\tau}\nabla_\sigma\nabla_\tau A^\rho = 0, \tag{2.8}$$

which is one of the conventional forms of equation for $A^\mu$. What is not conventional in the following discussion is the form of gauge condition which is consistent with the energy-momentum conservation law. The energy-momentum tensor $T^\nu_\mu$ of a free electromagnetic wave field is given by

$$T^\nu_\mu = -F_{\mu\sigma}F^{\nu\sigma}. \tag{2.9}$$

Since the divergence of (2.9) becomes

$$\nabla_\nu T^\nu_\mu = -\nabla_\nu(F_{\mu\sigma}F^{\nu\sigma}) = -F_{\mu\sigma}\nabla_\nu F^{\nu\sigma} = 0, \tag{2.10}$$

we see that assuming (2.8) is equivalent to

$$-F_{\mu\sigma}\nabla_\nu F^{\nu\sigma} = F_{\mu\sigma}g^{\sigma\tau}\nabla_\tau\phi = 0, \tag{2.11}$$

where $\phi \equiv \nabla_\rho A^\rho$ may be identified with Nakanishi's B-field [8]. Since $\nabla_\sigma\nabla_\nu F^{\nu\sigma}$ vanishes identically, we must have

$$g^{\sigma\tau}\nabla_\sigma\nabla_\tau\phi = 0. \tag{2.12}$$

Thus, we have shown that

$$\nabla_\nu F^{\nu\sigma} + g^{\sigma\tau}\nabla_\tau\phi = 0 \tag{2.13}$$

is not only mathematically consistent but physically relevant relation satisfying the energy-momentum conservation law.

The above arguments clearly suggest that a gauge fixing on $\nabla_\rho A^\rho$ should not be regarded as an auxiliary mathematical constraint to remove redundant degree of freedom but as an excited physical state of $\nabla_\nu A^\nu$ whose "ground state" is described by the non-divergent Lorentz gauge condition. In a conventional electromagnetic theory, any physically meaningful quantity is considered to be directly related with gauge invariance. In our new formulation, the same situation holds good if we only consider the restricted gauge transformation of the form:

$$\tilde{A}_\mu = A_\mu + \nabla_\mu\chi\ ; \quad g^{\mu\nu}\nabla_\mu\nabla_\nu\chi = 0, \tag{2.14}$$

which is closely related to the conservation of energy-momentum tensor through (2.10) and (2.12). Directly from (2.7), we see that both of the first and the second terms on the r.h.s. are invariant under the above gauge transformation.

If we accept that $\nabla_\nu A^\nu$ is a physical quantity, then it must be treated on equal footing with $F^{\mu\nu}$, which is realized in the following Lagrangian approach. Consider a Lagrangian density of the form:

$$\mathcal{L}^* = \mathcal{L} + \mathcal{L}_{GF} = -\frac{1}{4}F_{\mu\nu}F^{\mu\nu} - \frac{1}{2}(\nabla_\nu A^\nu)^2, \tag{2.15}$$

then, from which variation with respect to $A_\nu$, we readily recover (2.13), namely,

$$(\nabla_\mu F^{\mu\nu} + g^{\nu\mu}\nabla_\mu\phi)\delta A_\nu = 0. \tag{2.16}$$

Although the second term in (2.15) is called as a gauge fixing (GF) term while the first term is gauge-invariant, in our new formulation, both of them are "gauge invariant" in the sense of (2.14). Our classical way of introducing (2.15) based on the conservation of energy-momentum tensor is consistent with Nakanish-Lautrup (NL) formalism [6] for gauge field quantization in which $\nabla_\mu A^\mu$ plays a key role for the equation of propagator to have meaningful solutions. In NL formalism, GF Lagrangian density is given by

$$\mathcal{L}_{GF} = B\nabla_\mu A^\mu + \frac{\alpha}{2}B^2, \tag{2.17}$$

where $B$ field satisfies:

$$\nabla_\mu A^\mu + \alpha B = 0; \quad g^{\sigma\tau}\nabla_\sigma\nabla_\tau B = 0. \tag{2.18}$$

By comparing (2.15) with [(2.17), (2.18)], we get Feynman gauge of $\alpha = 1$.

## 2.3 Physical Interpretation of Gauge Condition

The vector equation (2.8) can be considered as a balance equation between the rotational ($\alpha^\rho$) and irrotational ($\nabla^\rho\chi$) parts of $A^\rho$ where $A^\rho = \alpha^\rho + \nabla^\rho\chi$. For simplicity, we consider the balance in a flat Minkowski space $\{x^\mu\}$ where the operator $\nabla_\mu$ is replaced by $\partial_\mu$ and $R^\rho_\sigma = 0$. Under this assumption, (2.8) is rewritten as

$$\partial^\tau\partial_\tau\alpha_\rho + \partial_\rho(\partial_\tau A^\tau) = 0. \tag{2.19}$$

For a given non-zero $\partial_\tau A^\tau$, the solution $\alpha_\rho$ to (2.19) consists of a homogeneous solution satisfying $\partial^\tau\partial_\tau\alpha^{(h)}_\rho = 0$ which corresponds to transverse modes and an inhomogeneous one $\alpha^{(i)}_\rho$ given by

$$\partial^\tau \partial_\tau \alpha_\rho^{(i)} + \partial_\rho(\partial_\tau A^\tau) = 0. \tag{2.20}$$

An example of such a balance equation is well-known for two dimensional irrotational motion of an incompressible fluid. Let $(u, v)$ be a Cartesian velocity field of the fluid. The incompressibility of fluid means that its motion is non-divergent: $\partial_x u + \partial_y v = 0$, while the vorticity must vanish since its motion is irrotational, i.e., $\partial_x v - \partial_y u = 0$. As a result of these two constraints, the velocity field $(u, v)$ must have a dual expression,

$$u = \partial_x \hat{\phi} = \partial_y \hat{\psi}; \quad v = \partial_y \hat{\phi} = -\partial_x \hat{\psi}, \tag{2.21}$$

where $\hat{\phi}$ and $\hat{\psi}$ are velocity potential and streamfunction, respectively and (2.21) is equivalent to the Cauchy–Riemann equations, so that $\hat{\phi}$ and $\hat{\psi}$ are the real and imaginary parts of a certain analytic function satisfying $\nabla^2\hat{\phi} = 0$ and $\nabla^2\hat{\psi} = 0$. In addition, since $\partial_x \hat{\phi} \partial_x \hat{\psi} + \partial_y \hat{\phi} \partial_y \hat{\psi} = 0$, they are called conjugate harmonics.

In what follows, we will show that, owing to (2.20) and a relation similar to (2.21), $\partial_\tau A^\tau$ can be related to a longitudinal electric field. In (2.20), we can assume without loss of generality that non-zero components are those with subscript $\rho$ taking values $\rho = 0$ and $\rho = 1$. Corresponding to this specification of propagation direction, we can express $\partial_\tau A^\tau$ as

$$\partial_\tau A^\tau = \partial_0 A^0 + \partial_1 A^1. \tag{2.22}$$

Redefining $\beta_\rho \equiv \partial^\tau \partial_\tau \alpha_\rho^{(i)}$ and $\xi \equiv \partial_\tau A^\tau$, (2.20) is rewritten as

$$\beta_\rho + \partial_\rho \xi = 0. \tag{2.23}$$

Since $\beta^\rho$ is divergence free by definition, i.e., $\partial_\rho \beta^\rho = 0$, there exists $\varpi$ satisfying

$$\beta^0 = \beta_0 = \partial_1 \varpi; \quad \beta^1 = -\beta_1 = -\partial_0 \varpi. \tag{2.24}$$

Combining (2.23) and (2.24), we obtain

$$\partial_1 \varpi + \partial_0 \xi = 0, \quad \partial_0 \varpi + \partial_1 \xi = 0. \tag{2.25}$$

Note that $F^{\nu\sigma}$ in Maxwell equation $j^\sigma = \partial_\nu F^{\nu\sigma}$ can be considered mathematically as a sort of extended "streamfunctions" for non-divergent vector $j^\sigma$. Since the dimension of $\beta^\sigma$ is the same as that of $j^\sigma$, we can use this relation to redefine $\varpi$ in (2.24), whose consistency in the present context will be checked shortly. Thus, we have

$$\beta^0 = \partial_1(-F^{01}); \quad \beta^1 = \partial_0(-F^{10}), \tag{2.26}$$

where $j^\sigma = \partial_\nu(-F^{\nu\sigma})$ is used instead of $j^\sigma = \partial_\nu(F^{\nu\sigma})$, in view of the irrelevance of the sign in the equation to determine "streamfunctions", in contrast to the situation in Maxwell equation: the reason why we only use $F^{01}$ here is because $\xi = \partial_\tau A^\tau$

in (2.22) is expressed solely in terms of $A^0$ and $A^1$. Substitution of $\varpi = -F^{01} = F_{01} = \partial_0 A_1 - \partial_1 A_0$ and $\xi = \partial_0 A^0 + \partial_1 A^1 = \partial_0 A^0 - \partial_1 A_1$ into (2.25) yields

$$(\partial^2_{00} - \partial^2_{11})A_0 = 0; \quad (\partial^2_{00} - \partial^2_{11})A_1 = 0, \tag{2.27}$$

which is an expected property of $A_\sigma$ for propagating waves and justifies our choice of $\varpi = -F^{01}$. Similarly, by cross differentiating (2.25), we get

$$(\partial^2_{00} - \partial^2_{11})\varpi = 0; \quad (\partial^2_{00} - \partial^2_{11})\xi = 0. \tag{2.28}$$

For a simple plane wave solution of $A_0 = A_1 = \exp[i(kx^0 - kx^1)]$, we get $\varpi = \xi = 2ik\exp[i(kx^0 - kx^1)]$ which shows that $\partial_\tau A^\tau$ in this case is equal to the electric field $F_{01}$ of a longitudinal mode.

Concerning the physical reality of vector potential $A_\sigma$, what has actually been verified in the experiments performed by Tonomura et al. [9] is the significance of the rotational component of $A_\sigma$ and the irrotational one was left untouched. The analysis presented above, however, shows that the irrotational part of $A_\sigma$ also has a physical meaning and should not be treated as a simple mathematical condition. If we accept this, then Lorenz gauge condition $\partial_\tau A^\tau = 0$ can be regarded as a "ground" or unexcited state of the divergence of a physically meaningful vector $A_\sigma$.

## 2.4 Clebsch Dual Field for Electromagnetic Waves

### *2.4.1 Lightlike Case*

Although the wave-particle duality of a given field is considered as an exclusive feature of quantum field theory (QFT), we can say that this ubiquitous feature of physical fields is reflected in the classical theory of light in a fragmented fashion, namely, Maxwell's theory represents the wave aspect of light field while geometric optics captures the corpuscular aspect. Specifically, the latter aspects is described mathematically as a fact that light rays give null geodesics on 4d pseudo Riemannian manifold $\mathcal{M}$ defined in (2.5), from which we see that an irrotational vector field $U_\mu = \nabla_\mu \chi$ with $\nabla^\nu \chi \nabla_\nu \chi = 0$ automatically satisfies (2.5). It is unclear, however, that how this solution is related to Maxwell's equation in which bivector fields play a key role.

From the viewpoint of vortex dynamics, it can be shown that Clebsch parameterized (CP) vector field provides an expected bivector field which behaves like $F_{\mu\nu}$ for electromagnetic free waves. CP vector field $\hat{U}_\mu$ was originally introduced to study Hamiltonian structure [10] of $3d$ barotropic fluid for which it is given by

$$\hat{U}_\mu = U_\mu + \nabla_\mu \chi, \quad \text{where} \quad U_\mu \equiv \lambda \nabla_\mu \phi. \tag{2.29}$$

Note that the vorticity field $S_{\mu\nu}$ defined as

$$S_{\mu\nu} \equiv \nabla_\mu U_\nu - \nabla_\nu U_\mu, \tag{2.30}$$

is written in this case as a simple bivector of the form

$$S_{\mu\nu} = L_\mu C_\nu - L_\nu C_\mu, \quad \text{where} \quad L_\mu \equiv \nabla_\mu \lambda \quad \text{and} \quad C_\mu \equiv \nabla_\mu \phi, \tag{2.31}$$

namely, $S_{\mu\nu}$ can be written in terms of the exterior product of $L_\mu$ and $C_\nu$.

One of the direct consequences of (2.29) is the following identity:

$$S_{\mu[\nu} S_{\rho\sigma]} = 0 \tag{2.32}$$

which can be directly obtained by substituting $S_{\mu\nu}$ into (2.32) which is alternatively written as

$$D \equiv S_{01} S_{23} + S_{02} S_{31} + S_{03} S_{12} = 0, \tag{2.33}$$

where $D = Pf(S)$ is the Pfaffian of the anti-symmetric matrix $S_{\mu\nu}$:$D^2 = Det(S_{\mu\nu})$. Similarly to $\boldsymbol{E} \cdot \boldsymbol{M} = 0$ for the electromagnetic field $F_{\mu\nu}$, the orthogonality between $(S_{01}, S_{02}, S_{03})$ and $(S_{23}, S_{31}, S_{12})$ should hold as a necessary condition for the existence of non-zero vector $U^\nu$ satisfying the matrix equation $S_{\mu\nu} U^\nu = 0$ to be shown shortly (in (2.46)) because of $Det(S_{\mu\nu}) = D^2$. Then, the matrix elements $S_{\mu\nu}$ have $(6 - 1) = 5$ independent components, and hence, they can be viewed as the homogeneous coordinates of a projective space with dimensionality $(5 - 1) = 4$. We recall here, as has been noticed by Ojima [11], the well-known relations among projective space, Grassmannian manifolds and Plücker coordinates of the latter: while a Grassmannian manifold $GM(p, n; F)$ defined by the set of all linear subspaces with a fixed dimensionality $p$ in a given $n$-dimensional linear space $F^n$ over a scalar field $F$ is more general than the concept of a projective space $PF^q = GM(1, q + 1; F)$, the Grassmannian manifold can, however, be embedded in a higher dimensional projective space by means of the so-called Plücker coordinates consisting of minor determinants of an anti-symmetric matrix constrained by them. Adapting this viewpoint to the Grassmannian manifold $GM(2; 4)$ of all the 2-dimensional linear subspaces of $R^4$, we can regard the above orthogonality relation (2.33) as the Plücker condition among the Plücker coordinates given by the matrix elements $S_{\mu\nu}$ [11]. In this context, it is seen that the necessary and sufficient condition for holding (2.33) is given by the simplicity of the bivector $S_{\mu\nu}$ (see, for instance, Huggett and Tod [12]).

For our purpose, a particularly important class of solutions is d'Alembert equation for a massless field $\phi$:

$$g^{\mu\nu} \nabla_\mu \nabla_\nu \phi = 0; \quad g^{\mu\nu} \nabla_\mu \phi \nabla_\nu \phi = 0. \tag{2.34}$$

A couple of equations given in (2.34) are well-documented ones whose compatibility is readily checked for a plane wave of the form: $\phi = \exp(ik_\nu x^\nu)$ with $k^\nu k_\nu = 0$ in a flat spacetime. Equation (2.34) assumes that the solutions to d'Alembert equation for a massless field exist on a curved spacetime keeping the compatibility conditions unchanged. Using this $\phi$, we can introduce anew a lightlike CP flow field as

$$U_\mu = \lambda \nabla_\mu \phi; \quad \text{where} \quad g^{\nu\sigma} \nabla_\sigma \nabla_\nu \phi = \nabla^\nu \nabla_\nu \phi = 0, \tag{2.35}$$

and $\lambda$ is assumed to be determined by a spacelike Klein–Gordan (KG) equation

$$\nabla^\nu \nabla_\nu \lambda - (\kappa_0)^2 \lambda = 0, \tag{2.36}$$

with $\kappa_0$ being a suitable constant, whose justification will be given later by (2.45). Substitution of (2.35) and (2.36) into the left-hand side of (2.5) with the use of self-orthogonality condition in (2.34), we have

$$U^\nu \nabla_\nu U_\mu = -S_{\mu\nu}(\lambda C^\nu) = (C_\mu L_\nu - L_\mu C_\nu)(\lambda C^\nu) = (L_\nu C^\nu)\lambda C_\mu. \tag{2.37}$$

Recall that, in referring to Cauchy–Riemann equation on (2.21), we touched on the orthogonality condition $\partial_x \hat{\phi} \partial_x \hat{\psi} + \partial_y \hat{\phi} \partial_y \hat{\psi} = 0$ holding between conjugate harmonics of $\nabla^2 \hat{\phi} = 0$ and $\nabla^2 \hat{\psi} = 0$. This condition is a directional one that can be imposed on a couple of given conjugate harmonics. A similar situation occurs in free propagating electromagnetic waves in Minkowski space for which we have

$$\partial^\tau \partial_\tau \boldsymbol{E} = 0, \quad \partial^\tau \partial_\tau \boldsymbol{M} = 0, \quad \text{with } \boldsymbol{E} \perp \boldsymbol{M}. \tag{2.38}$$

In determining a directional constraint in our model, we specify $L_\mu$ such that

$$C^\nu \nabla_\nu L_\mu = 0, \tag{2.39}$$

which can be compared to the fact that $\boldsymbol{E}$ and $\boldsymbol{M}$ are advected along a null Poynting vector. Then, it follows immediately that we have

$$L^\mu (C^\nu \nabla_\nu L_\mu) = 0; \quad \Rightarrow \quad C^\nu \nabla_\nu (L^\mu L_\mu) = 0; \tag{2.40}$$

$$\begin{aligned} C^\mu (C^\nu \nabla_\nu L_\mu) = 0; \quad &\Rightarrow \quad C^\nu [\nabla_\nu (C^\mu L_\mu) - L_\mu \nabla_\nu C^\mu] \\ &= \quad C^\nu \nabla_\nu (C^\mu L_\mu) = 0. \end{aligned} \tag{2.41}$$

In deriving (2.41), the use has been made of $C^\nu \nabla_\nu C^\mu = 0$, which says that $C^\nu$ itself satisfies the null geodesic equation. Equation (2.41) says that the scalar quantity $C^\nu L_\nu$ is advected along a null geodesic so that it allows us to have the following orthogonality condition between the two vectors $L^\mu$ and $C^\mu$.

$$L_\nu C^\nu = 0. \tag{2.42}$$

Notice that, using an vector identity:

$$C^\nu \nabla_\nu L_\mu = C^\nu \nabla_\mu L_\nu = \nabla_\mu (C^\nu L_\nu) - L_\nu \nabla_\mu C^\nu, \tag{2.43}$$

and (2.42), we get

$$C^\nu \nabla_\nu L_\mu = -L_\nu \nabla_\mu C^\nu = -L^\nu \nabla_\mu C_\nu = -L^\nu \nabla_\nu C_\mu = 0. \tag{2.44}$$

So, we see that there exists a reciprocity between $C_\mu$ and $L_\mu$ concerning the advective relation in (2.44). Since any vector perpendicular to a given null vector is either the same null vector or a spacelike one, we can choose $L_\nu$ such that it is a spacelike vector satisfying

$$\rho \equiv -L^\nu L_\nu > 0, \tag{2.45}$$

which justifies the spacelike KG equation in (2.36). With the orthogonality condition (2.42), (2.37) becomes

$$U^\nu \nabla_\nu U_\mu = -S_{\mu\nu}(\lambda C^\nu) = -S_{\mu\nu} U^\nu = 0, \tag{2.46}$$

which says that CP flow (2.35) satisfies a null geodesic equation. It is worthwhile to point out a similarity between (2.46) and the following equation for electromagnetic field:

$$F_{\mu\nu} P^\nu = 0, \tag{2.47}$$

where $F_{\mu\nu}$ and $P^\nu$ denote, respectively, the electromagnetic bivector and Poynting 4-vector. The latter one is parallel to a null geodesic perpendicular to both the electric $\boldsymbol{E} = (F_{01}, F_{02}, F_{03})$ and the magnetic $\boldsymbol{B} = -(F_{23}, F_{31}, F_{12})$ bivectors. The difference in form between (2.46) and (2.47) is that the bivector $F_{\mu\nu}$ is not defined as the curl of $P_\mu$ while $S_{\mu\nu}$ is derived by the curl of $U_\mu$.

The energy-momentum tensor $T_\mu^\nu$ associated with electromagnetic radiation field is given in its mixed tensor form by

$$T_\mu^\nu = -F_{\mu\sigma} F^{\nu\sigma}. \tag{2.48}$$

The corresponding quantity $\hat{T}_\mu^\nu$ for $S_{\mu\nu}$ is given by

$$\begin{aligned} \hat{T}_\mu^\nu &= -S_{\mu\sigma} S^{\nu\sigma} = -(L_\mu C_\sigma - C_\mu L_\sigma)(L^\nu C^\sigma - C^\nu L^\sigma) \\ &= -(L_\sigma L^\sigma) C_\mu C^\nu = \rho C_\mu C^\nu, \end{aligned} \tag{2.49}$$

by the use of (2.34), (2.42) and (2.45). Though $C^\mu$ is lightlike, (2.49) is identical in form to the energy-momentum tensor of free moving (fluid) particles. Actually, following from $C^\nu \nabla_\nu C_\mu = 0$, $\nabla_\nu C^\nu = 0$ and $C^\nu \nabla_\nu \rho = 0$, which are derived by (2.34), (2.39) and (2.45), we get

$$\nabla_\nu \hat{T}_\mu^\nu = 0. \tag{2.50}$$

Since (2.49) has dual representations, (2.50) can also be expressed in terms of $S_{\mu\nu}$ as

$$\nabla_\nu \hat{T}_\mu^\nu = -\nabla_\nu (S_{\mu\sigma} S^{\nu\sigma}) = -S_{\mu\sigma} \nabla_\nu S^{\nu\sigma} = 0, \tag{2.51}$$

where the uses have been made of $\nabla_\nu S_{\rho\sigma} + \nabla_\rho S_{\sigma\nu} + \nabla_\sigma S_{\nu\rho} = 0$ and $S_{\nu\rho}S^{\nu\rho} = 0$ corresponding to $F_{\nu\rho}F^{\nu\rho} = 0$ in the original equation. In Maxwell's electromagnetic theory $\nabla_\nu T^\nu_\mu = -F_{\mu\sigma}\nabla_\nu F^{\nu\sigma}$ vanishes under the following condition of no electric current:

$$\nabla_\nu F^{\nu\sigma} = 0, \tag{2.52}$$

which yields an electromagnetic wave equation in the vacuum. Notice, however, that $\nabla_\nu S^{\nu\sigma} = 0$ is a sufficient condition for (2.51) but not a necessary one. Directly from the definition of $S_{\mu\nu}$, namely, (2.31), and using (2.44), (2.34) we have

$$\nabla_\nu S^{\nu\sigma} = (\nabla_\nu L^\nu)C^\sigma + [L^\nu\nabla_\nu C^\sigma - C^\nu\nabla_\nu L^\sigma] - (\nabla_\nu C^\nu)L^\sigma = (\nabla_\nu L^\nu)C^\sigma. \tag{2.53}$$

By similar manipulations, we also get

$$C^\sigma\nabla_\sigma S_{\mu\nu} = C^\sigma\nabla_\sigma(L_\mu C_\nu - L_\nu C_\mu) = 0. \tag{2.54}$$

So, we see that $S_{\mu\nu}$ is advected by null vector $C^\sigma$. Note that the second equation in (2.34) and (2.42) are rewritten, respectively, as

$$C^\nu\nabla_\nu\phi = 0; \quad C^\nu\nabla_\nu\lambda = 0. \tag{2.55}$$

In hydrodynamic terms, (2.54) tells us that, in addition to $\phi$ and $\lambda$, vorticity $S_{\mu\nu}$ is also advected (or convected) along a null geodesic with tangent vector $C^\mu$. Since the CP vortex dynamics is characterised by a couple of variables $\phi$ and $\lambda$, (2.54) and (2.55) suggest that $S_{\mu\nu}$ can be parameterized in terms of co-moving Lagrange coordinates $\phi$ and $\lambda$ as

$$S_{\mu\nu} = S_{\mu\nu}(\phi, \lambda). \tag{2.56}$$

Polarization is an important aspect of electromagnetic waves which is related to spin dynamics, as right and left circularly polarized states describe spin states of a photon. To see how polarization is represented in our formulation, we first note that the formulation on CP vortex dynamics can possess a pair of parameter spaces of $[(\lambda, \phi); ({}^\star\lambda, {}^\star\phi(\phi))]$ illustrated in the following Fig. 2.1 where, without loss of generality, the spatial part of covector $C_\mu$ at the point of interest is assumed to be parallel to $x^1$ axis and that of covector $L_\mu$ lies on the plane spanned by the two coordinate axes $x^1$ and $x^3$. In such a configuration, the magnetic counterpart vector $\boldsymbol{M}_{(s)}$ in $S_{\mu\nu}$ system becomes parallel (or anti-parallel) to $x^2$ axis. The $\star$ quantities corresponding to $C_\mu$ and $L_\mu$ together with the associated definition of bivector and orthogonality condition are respectively introduced as follows:

$${}^\star L_\mu \equiv \nabla_\mu({}^\star\lambda); \quad {}^\star C_\mu \equiv \nabla_\mu{}^\star\phi(\phi) = \varpi(\phi)C_\mu, \tag{2.57}$$

$${}^\star S_{\mu\nu} \equiv {}^\star C_\mu({}^\star L_\nu) - ({}^\star L_\mu){}^\star C_\nu, \quad {}^\star L_\nu({}^\star C^\nu) = 0, \tag{2.58}$$

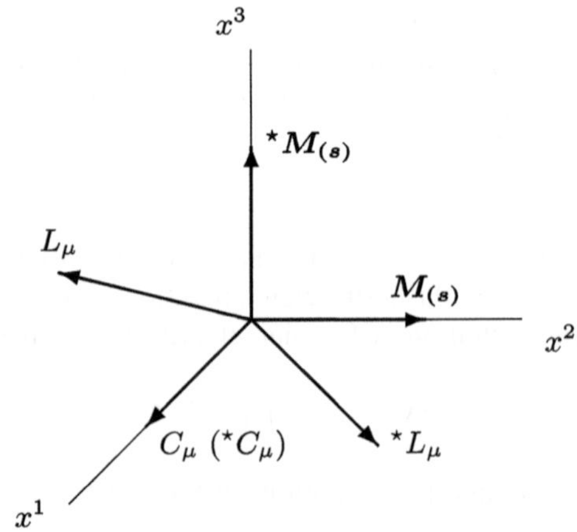


**Fig. 2.1** Dual configuration of $L_\mu$ and $^*L_\mu$ and associated $\boldsymbol{M}_{(s)}$ and $^\star\boldsymbol{M}_{(s)}$

where $\varpi(\phi) \equiv d(^\star\phi)/d\phi$. The important points illustrated in Fig. 2.1 are that the spatial part of $^\star C_\mu$ is parallel (or anti-parallel) to $C_\mu$ while $^\star L_\mu$ lies not on the plane spanned by $x^1$ and $x^3$ but on the one spanned by $x^1$ and $x^2$, from which we see that $^\star\boldsymbol{M}_{(s)}$ becomes parallel (or anti-parallel) to $x^3$ axis. A little bit lengthy but straightforward calculations using (2.34), (2.42), (2.57) and (2.58) show that the orthogonality condition holding between $\boldsymbol{M}_{(s)}$ and $^\star\boldsymbol{M}_{(s)}$ is equivalent to the one between $L_\mu$ and $^\star L_\mu$, namely,

$$S_{23}(^\star S_{23}) + S_{31}(^\star S_{31}) + S_{12}(^\star S_{12}) = -\varpi(\phi)(C_0)^2 L_\nu(^\star L^\nu) = 0. \tag{2.59}$$

Now, consider simple examples in a flat spacetime: a solution $S_{\mu\nu}$ of the form:

$$\lambda = \sin\theta_\lambda, \quad \phi = \sin\theta_\phi; \quad \theta_\lambda \equiv kx^0 - kx^1 + lx^3, \quad \theta_\phi \equiv kx^0 - kx^1, \tag{2.60}$$

where $k$ and $l$ are two positive constants. From (2.31), we get the following non-zero components of $S_{\mu\nu}$:

$$E^3_{(s)} = S_{03} = -kl\cos\theta_\lambda\cos\theta_\phi; \quad M^2_{(s)} = S_{31} = -kl\cos\theta_\lambda\cos\theta_\phi. \tag{2.61}$$

Clearly (2.61) is a linearly polarized wave and the spatial part of covector $L_\mu$ lies in the plane spanned by $C_1$ and $E^3_{(s)}$. In order to have a circularly polarized wave, we need an additional linearly polarized one to superimpose on (2.61). An example of this additional mode denoted by $(^\star\boldsymbol{E}_{(s)}, ^\star\boldsymbol{M}_{(s)})$ is given by, say,

$$^\star\lambda = \cos\hat{\theta}_\lambda, \quad ^\star\phi = \sqrt{1-\phi^2} = \cos\theta_\phi; \quad \hat{\theta}_\lambda \equiv kx^0 - kx^1 + lx^2, \tag{2.62}$$

which yields

$$^\star E^2_{(s)} =^\star S_{02} = -kl\sin\hat{\theta}_\lambda\sin\theta_\phi; \quad ^\star M^3_{(s)} =^\star S_{12} = kl\sin\hat{\theta}_\lambda\sin\theta_\phi, \tag{2.63}$$

where, as already pointed out, ${}^\star C_\mu$ is parallel (or anti-parallel) to $C_\mu$ and the spatial part of ${}^\star L_\mu$ is now not in the plane spanned by $C_1$ and $E^3_{(s)}$ but in the plane spanned by $C_1$ and $M^2_{(s)}$. As is seen in the examples above, the four vectors $C^\mu$, $L^\mu$, ${}^\star C^\mu$ and ${}^\star L^\mu$, satisfying the following mutually orthogonal conditions (2.64) directly derived from (2.42), (2.57), (2.58) and (2.59), serve as the complete bases of $4d$ spacetime with which circularly polarized states can be constructed.

$$\begin{aligned} C_\nu {}^\star C^\nu = 0, \quad L_\nu {}^\star L^\nu = 0, \\ C_\nu L^\nu = 0, \quad C_\nu {}^\star L^\nu = 0, \\ {}^\star C_\nu L^\nu = 0, \quad {}^\star C_\nu {}^\star L^\nu = 0. \end{aligned} \tag{2.64}$$

It is worthwhile to point out that similar orthogonality conditions also hold for the bivector components $S_{\mu\nu}$ if we consider them as the components of dual spinor representations of the following "Poynting" 4-vector $\Pi^\mu$ that can be derived from (2.49), namely,

$$\Pi^0 \equiv (S_{23})^2 + (S_{31})^2 + (S_{12})^2; \quad \Pi^1 \equiv -(S_{02}S_{12} - S_{03}S_{31}); \tag{2.65}$$

$$\Pi^2 \equiv -(S_{03}S_{23} - S_{01}S_{12}); \quad \Pi^3 \equiv -(S_{01}S_{31} - S_{02}S_{23}), \tag{2.66}$$

where the sign of spatial components of $\Pi^\mu$ is chosen identically to that of the metric $diag(1, -1, -1, -1)$ referred to at the beginning of Sect. 2.2. To see this, for simplicity, let us consider a linearly polarized wave whose $\Pi^\mu$, $\boldsymbol{E}_{(s)}$ and $\boldsymbol{M}_{(s)}$ are oriented such that they are parallel (or anti-parallel), respectively, to the orthogonal axes $x^1$, $x^2$ and $x^3$ of a Lorentz reference frame. Equations (2.62) and (2.63) serve as an example for this configuration. Since $E^2_{(s)} = S_{02}$ and $M^3_{(s)} = S_{12}$, using (2.65), we have $\Pi^1 = -S_{02}S_{12}$. Substituting this into the following definition of spinor representation of $\Pi^\mu$ [11], we obtain

$$\sqrt{2}\Psi(\Pi^\mu) \equiv \begin{pmatrix} \Pi^0 + \Pi^3 & \Pi^1 + i\Pi^2 \\ \Pi^1 - i\Pi^2 & \Pi^0 - \Pi^3 \end{pmatrix} = \begin{pmatrix} \Pi^0 & \Pi^1 \\ \Pi^1 & \Pi^0 \end{pmatrix}. \tag{2.67}$$

The reason why we use real variables in the above lightlike modes is just for the sake of simplicity and full complexification of variables is to be introduced for the discussion of non-lightlike modes in the following subsection. So, we think that the usage of complex spinor representation like (2.67) in our present discussion is not superficial though it may look so as far as we stay in the restricted representation by real variables. Using $(S_{02})^2 = (S_{12})^2$, we get

$$\sqrt{2}\Psi(\Pi^\mu) = \begin{pmatrix} (S_{02})^2 & -S_{02}S_{12} \\ -S_{02}S_{12} & (S_{02})^2 \end{pmatrix}, \tag{2.68}$$

which becomes equal to the non-zero $2 \times 2$ minor matrix of degenerated $\hat{T}^{\mu\nu}$: $\hat{T}^{MN}_{(m)}$; $0 \le M, N \le 1$ (cf. (2.49 )). Equating (2.68) with $\hat{T}^{MN}_{(m)}$, we have

$$\sqrt{2}\Psi(\Pi^{\mu}) = \begin{pmatrix} (S_{02})^2 & -S_{02}S_{12} \\ -S_{02}S_{12} & (S_{12})^2 \end{pmatrix} = \begin{pmatrix} \sqrt{\rho}C^0\sqrt{\rho}C^0 & \sqrt{\rho}C^0\sqrt{\rho}C^1 \\ \sqrt{\rho}C^1\sqrt{\rho}C^0 & \sqrt{\rho}C^1\sqrt{\rho}C^1 \end{pmatrix}. \tag{2.69}$$

So, using (2.49), we see that the spinor representation of $\Pi^{\mu}$ also has a dual form in which no undetermined parameter is involved and that $(\sqrt{\rho}C^0, \sqrt{\rho}C^1, S_{02}, S_{12})$ gives a complete orthogonal set of spinor components with which not only complementary aspects (wave vs. particle) of the radiation field (2.49) can be described, but also a set of orthogonal bases of spacetime is provided. A more noteworthy aspect on this duality is the fact that $S_{\mu\nu}$ is not only endowed with the property of the orthogonal bases of the spacetime but also with that of curvature of the spacetime, which will be shown in what follows. Apparently, two vortex dynamics $F_{\mu\nu}$ and $S_{\mu\nu}$ have seemingly different origins. Nevertheless, as we have just shown here, there exists a remarkable similarity between them, which implies a possibility that the newly introduced latter field may be the extended field of the former from a group theoretical point of view.

The Riemann curvature tensor satisfies the following properties:

$$R_{\mu\nu\rho\sigma} = R_{[\mu\nu]\rho\sigma} = R_{\mu\nu[\rho\sigma]};\ R_{\mu\nu\rho\sigma} = R_{\rho\sigma\mu\nu}; \tag{2.70}$$

$$R_{\mu[\nu\rho\sigma]} = 0;\ \nabla_{[\mu}R_{\nu\rho]\sigma\tau} = 0. \tag{2.71}$$

The last two equalities in (2.71) are called the first and the second Bianchi identities. If we introduce $\hat{S}_{\mu\nu\rho\sigma} = S_{\mu\nu}S_{\rho\sigma}$, then using (2.32), we readily see that it satisfies:

$$\hat{S}_{\mu\nu\rho\sigma} = \hat{S}_{[\mu\nu]\rho\sigma} = \hat{S}_{\mu\nu[\rho\sigma]};\quad \hat{S}_{\mu\nu\rho\sigma} = \hat{S}_{\rho\sigma\mu\nu};\quad \hat{S}_{\mu[\nu\rho\sigma]} = 0. \tag{2.72}$$

As to the property corresponding to the second Bianchi identity, we first calculate the quantity: $J_{\mu\nu\rho\sigma\tau} = \nabla_{\mu}\hat{S}_{\nu\rho\sigma\tau} + \nabla_{\nu}\hat{S}_{\rho\mu\sigma\tau} + \nabla_{\rho}\hat{S}_{\mu\nu\sigma\tau}$;

$$\begin{aligned} J_{\mu\nu\rho\sigma\tau} &= \nabla_{\mu}\hat{S}_{\nu\rho\sigma\tau} + \nabla_{\nu}\hat{S}_{\rho\mu\sigma\tau} + \nabla_{\rho}\hat{S}_{\mu\nu\sigma\tau} \\ &= (\nabla_{\mu}S_{\nu\rho} + \nabla_{\nu}S_{\rho\mu} + \nabla_{\rho}S_{\mu\nu})S_{\sigma\tau} + (S_{\nu\rho}\nabla_{\mu} + S_{\rho\mu}\nabla_{\nu} + S_{\mu\nu}\nabla_{\rho})S_{\sigma\tau} \\ &= (S_{\nu\rho}\nabla_{\mu} + S_{\rho\mu}\nabla_{\nu} + S_{\mu\nu}\nabla_{\rho})S_{\sigma\tau} \end{aligned} \tag{2.73}$$

where the use has been made of an identity: $\nabla_{\mu}S_{\nu\rho} + \nabla_{\nu}S_{\rho\mu} + \nabla_{\rho}S_{\mu\nu} = 0$. Secondly, referring to (2.3), consider the Hodge dual of $S^{\mu\nu}$. In a local Lorentz reference frame, its components ${}^*S^{\mu\nu}_{(L)}$ can be rewritten in terms of those of $S^{(L)}_{\mu\nu}$. If we write down its components, we get

$$({}^*S^{\mu\nu}_{(L)}) = \begin{pmatrix} 0 & S^{(L)}_{23} & S^{(L)}_{31} & S^{(L)}_{12} \\ -S^{(L)}_{23} & 0 & S^{(L)}_{03} & -S^{(L)}_{02} \\ -S^{(L)}_{31} & -S^{(L)}_{03} & 0 & S^{(L)}_{01} \\ -S^{(L)}_{12} & S^{(L)}_{02} & -S^{(L)}_{01} & 0 \end{pmatrix}. \tag{2.74}$$

By direct calculation, we obtain

$$(^*S^{\mu\nu}_{(L)}\partial_\nu\theta) = \begin{pmatrix} +(S^{(L)}_{23}\partial_1\theta + S^{(L)}_{31}\partial_2\theta + S^{(L)}_{12}\partial_3\theta) \\ -(S^{(L)}_{23}\partial_0\theta + S^{(L)}_{30}\partial_2\theta + S^{(L)}_{02}\partial_3\theta) \\ -(S^{(L)}_{31}\partial_0\theta + S^{(L)}_{10}\partial_3\theta + S^{(L)}_{03}\partial_1\theta) \\ -(S^{(L)}_{12}\partial_0\theta + S^{(L)}_{20}\partial_1\theta + S^{(L)}_{01}\partial_2\theta) \end{pmatrix}, \tag{2.75}$$

whose every right-hand side component has the form of $\pm(S^{(L)}_{\nu\rho}\partial_\mu + S^{(L)}_{\rho\mu}\partial_\nu + S^{(L)}_{\mu\nu}\partial_\rho)\theta$. Using the definition of $S_{\mu\nu}$ in (2.30) and substituting $\lambda$ and $\phi$ into $\theta$ in the above expression, we get

$$(S^{(L)}_{\nu\rho}\partial_\mu + S^{(L)}_{\rho\mu}\partial_\nu + S^{(L)}_{\mu\nu}\partial_\rho)\phi = 0;\ (S^{(L)}_{\nu\rho}\partial_\mu + S^{(L)}_{\rho\mu}\partial_\nu + S^{(L)}_{\mu\nu}\partial_\rho)\lambda = 0. \tag{2.76}$$

On the other hand, in the local Lorentz reference frame, (2.73) assumes the form

$$J_{\mu\nu\rho\sigma\tau} = (S^{(L)}_{\nu\rho}\partial_\mu + S^{(L)}_{\rho\mu}\partial_\nu + S^{(L)}_{\mu\nu}\partial_\rho)S^{(L)}_{\sigma\tau}. \tag{2.77}$$

So substituting (2.56) into (2.77) and using (2.76), we finally get $J_{\mu\nu\rho\sigma\tau} = 0$. Since it is a tensor quantity, it must vanish in any other coordinate systems. Thus we get

$$\nabla_{[\mu}\hat{S}_{\nu\rho]\sigma\tau} = 0. \tag{2.78}$$

### 2.4.2 Non-lightlike Case

The energy-momentum tensor of CP flow field given by (2.49) corresponds to a case of lightlike flow. In this subsection, we show that the arguments on lightlike case are readily extended to non-lightlike cases. In what follows, we assume that the dynamical variables involved in CP formalism are complexified and start from an essentially the same but a slightly different form of the definition of CP flow field:

$$U_\mu = \frac{1}{2}(\lambda C_\mu - \phi L_\mu). \tag{2.79}$$

As in the case of lightlike flow, we assume that $U_\mu$ satisfies the geodesic equation (2.5) now retaining non-zero $V \equiv U^\nu U_\nu/2$, namely,

$$-S_{\mu\nu}U^\nu + \nabla_\mu V = 0. \tag{2.80}$$

In (2.6), we referred to the normalization of velocity four vector $U^\mu$ in which its magnitude $V$ becomes either 0 (for lightlike velocity) or $\pm 1$ (for timelike/spacelike one). The magnitude of a lightlike vector $C^\mu = \nabla^\mu\phi$ where we have $\nabla^\nu\nabla_\nu\phi = 0$ discussed in Sect. 2.4.1 is given by the second equation in (2.34), namely, $C^\nu C_\nu = \nabla^\nu\phi\nabla_\nu\phi = 0$. The extension of a scalar lightlike wave equation: $\nabla^\nu\nabla_\nu\phi = 0$ to a non-lightlike case is known as KG equation

$$g^{\mu\nu}\nabla_\mu\nabla_\nu\psi + m^2\psi = 0, \tag{2.81}$$

where $m$ is either a real or pure imaginary number to be discussed later. A complex plane wave solution to (2.81) having the form of $\psi = \exp i(k_\sigma x^\sigma)$ satisfies the following a couple of equations:

$$g^{\mu\nu}\nabla_\mu\nabla_\nu\psi = -k^\sigma k_\sigma\psi, \quad g^{\mu\nu}\nabla_\mu\psi\nabla_\nu\psi = -k^\sigma k_\sigma\psi^2, \tag{2.82}$$

which is a natural extension of (2.34). So, here we assume that a couple of (complex) Clebsch variables $\lambda$ and $\phi$ satisfy

$$g^{\mu\nu}\nabla_\mu\nabla_\nu\psi = -m^2\psi; \quad g^{\mu\nu}\nabla_\mu\psi\nabla_\nu\psi = -m^2\psi^2, \tag{2.83}$$

where $m$ for $\lambda$ may be different from that of $\phi$. It must be emphasized here that the compatibility of a couple of equations given in (2.83) depends exclusively on the assumption that a solution has a form of plane wave: $\psi = \exp i(k_\sigma x^\sigma)$. So, the results in what follows in this subsection also depend on this restriction of which meaning will be discussed in Sect. 2.5 from the viewpoint of off-shell dynamics of DPs as virtual particles.

For the plane wave solution mentioned above, since we have

$$g^{\mu\nu}\nabla_\mu\psi\nabla_\nu\psi^* = k^\sigma k_\sigma\psi\psi^*, \tag{2.84}$$

where $\psi^*$ denotes the complex conjugate of $\psi$, equating $k^\sigma k_\sigma$ with $m^2$, we can use

$$g^{\mu\nu}\nabla_\mu\psi\nabla_\nu\psi^* = m^2\psi\psi^*, \tag{2.85}$$

as the equation defining whether a given complex covector $\nabla_\mu\psi$ is timelike or not. Namely, since $\psi\psi^*$ is non negative, we can say that $\nabla_\mu\psi$ is timelike if $m^2 > 0$ and it is spacelike if $m^2 < 0$, which corresponds to the second equation in (2.6). Substitution of $\lambda$ and $\phi$ into (2.81) yields

$$\nabla_\nu C^\nu + (m_\phi)^2\phi = 0; \quad \nabla_\nu L^\nu + (m_\lambda)^2\lambda = 0, \tag{2.86}$$

and the second equation in (2.82) for $\phi$ and $\lambda$ respectively becomes

$$C^\nu C_\nu + (m_\phi)^2\phi^2 = 0; \quad L^\nu L_\nu + (m_\lambda)^2\lambda^2 = 0. \tag{2.87}$$

Note that neither the first nor the second equations in (2.82) give a directional constraint on $C_\mu$ and $L_\mu$ so that, in addition to (2.82), we can impose on them the important orthogonality constraint already given in (2.42), but in our present case, it must be defined in a complex form. Defining $C^\nu = d^\nu + ie^\nu$ and $L_\nu = p_\nu + iq_\nu$, we get

$$C^\nu L_\nu = (d^\nu p_\nu - e^\nu q_\nu) + i(d^\nu q_\nu + e^\nu p_\nu) = 0. \tag{2.88}$$

In $4d$ spacetime, it is always possible that we can specify the orientation of four vector $d^{\mu}$, $e^{\nu}$, $p^{\mu}$ and $q^{\mu}$ such that $d^{\nu}p_{\nu}=0$, $e^{\nu}q_{\nu}=0$, $d^{\nu}q_{\nu}=0$ and $e^{\nu}p_{\nu}=0$, which can be concisely rewritten as

$$C^{\nu}L_{\nu}=0; \quad C^{\nu}L^{*}_{\nu}=0, \tag{2.89}$$

where $L^{*}_{\nu}$ is the complex conjugate of $L_{\nu}$. With this orthogonality condition, it can be readily shown that $U^{\nu}$ is divergence free, namely, $\nabla_{\nu}U^{\nu}=0$, and using (2.87) and (2.89), $V$ now becomes

$$V=\left(\frac{1}{2}\right)^{3}(\lambda C^{\nu}-\phi L^{\nu})(\lambda C_{\nu}-\phi L_{\nu})=-\left(\frac{1}{2}\right)^{3}[(m_{\lambda})^{2}+(m_{\phi})^{2}](\lambda\phi)^{2}. \tag{2.90}$$

Now, going back to (2.80), direct calculations of $S_{\mu\nu}U^{\nu}$ and $\nabla_{\mu}V$ yield

$$-S_{\mu\nu}U^{\nu}+\nabla_{\mu}V=\frac{1}{2}[(m_{\lambda})^{2}-(m_{\phi})^{2}]\lambda\phi U_{\mu}, \tag{2.91}$$

so that (2.80) holds good only when $(m_{\lambda})^{2}=(m_{\phi})^{2}=m^{2}$. Under this condition, we also get

$$U^{\sigma}\nabla_{\sigma}(\lambda\phi)=0; \quad \Omega\equiv S_{\mu\nu}S^{\mu\nu}=2m^{4}(\lambda\phi)^{2}=-8m^{2}V, \tag{2.92}$$

from which we obtain an important advection equation:

$$U^{\sigma}\nabla_{\sigma}\Omega=0. \tag{2.93}$$

In Sect. 2.4.1, we have looked into the form of energy-momentum tensor given by (2.49). For non-lightlike case, if we follow the conventional electromagnetic knowledge again, it is natural to start with the form:

$$\hat{T}^{\nu}_{\mu}=-S_{\mu\sigma}S^{\nu\sigma}+\frac{1}{4}S_{\alpha\beta}S^{\alpha\beta}g^{\nu}_{\mu}. \tag{2.94}$$

Through well-known manipulations in electromagnetic theory, we get

$$\nabla_{\nu}\hat{T}^{\nu}_{\mu}=-S_{\mu\sigma}\nabla_{\nu}S^{\nu\sigma}. \tag{2.95}$$

By quite similar manipulations deriving (2.53), we have

$$\nabla_{\nu}S^{\nu\sigma}=(\nabla_{\nu}L^{\nu})C^{\sigma}-(\nabla_{\nu}C^{\nu})L^{\sigma}. \tag{2.96}$$

Therefore, (2.95) becomes

$$\begin{aligned}\nabla_{\nu}\hat{T}^{\nu}_{\mu}&=-S_{\mu\sigma}[(\nabla_{\nu}L^{\nu})C^{\sigma}-(\nabla_{\nu}C^{\nu})L^{\sigma}]\\&=-m^{2}S_{\mu\sigma}(-\lambda C^{\sigma}+\phi L^{\sigma})=2m^{2}S_{\mu\sigma}U^{\sigma}.\end{aligned} \tag{2.97}$$

Using (2.80), (2.90) and the second equation in (2.92), the above leads to

$$\nabla_\nu \hat{T}^\nu_\mu = \nabla_\nu \left( -\frac{1}{4}\Omega g^\nu_\mu \right). \tag{2.98}$$

Combining (2.94) and (2.98) together with the notation $\hat{S}_{\mu\nu\sigma\rho} = S_{\mu\nu}S_{\sigma\rho}$ introduced used in the previous Sect. 2.4.1, we finally obtain

$$\nabla_\nu \hat{G}^\nu_\mu = 0; \quad \hat{G}^\nu_\mu \equiv -\hat{S}_{\mu\sigma}{}^{\nu\sigma} + \frac{1}{2}\hat{S}_{\alpha\beta}{}^{\alpha\beta} g^\nu_\mu, \tag{2.99}$$

which tells that, as are shown in (2.72) and (2.78), $\hat{S}_{\mu\sigma}{}^{\nu\sigma}$ behaves like Riemann curvature tensors and hence (2.99) becomes isomorphic to Einstein equation:

$$\nabla_\nu G^\nu_\mu = 0; \quad G^\nu_\mu \equiv -R_{\mu\sigma}{}^{\nu\sigma} + \frac{1}{2}R_{\alpha\beta}{}^{\alpha\beta} g^\nu_\mu, \tag{2.100}$$

where $R_{\mu\nu\sigma\rho}$ denotes Riemann tensor.

In the previous Sect. 2.4.1 covering lightlike CP flows, we have shown that one of Clebsch variables $\lambda$ must satisfy spacelike KG equation (2.36). Using it, (2.53) can be rewritten as a spacelike "Proca equation" of the form

$$\nabla^\nu \nabla_\nu U^\sigma - (\kappa_0)^2 U^\sigma = 0, \tag{2.101}$$

though $U^\sigma$ is a null vector field. Since the original Proca equation assumes the form

$$\nabla^\nu \nabla_\nu A^\sigma + m^2 A^\sigma = 0, \quad m^2 > 0, \tag{2.102}$$

we call $U^\sigma$ satisfying (2.101) *Clebsch dual (CD) (electromagnetic wave) field* or CD field for short. In view of this property of CD flows, it is natural to consider spacelike vector $U^\sigma$ in formulating a non-lightlike case of CP flow dynamics. Based on (2.90), we can evaluate the physical magnitude of $V$ denoted as $\parallel V \parallel$ by

$$\parallel V \parallel = \frac{1}{8}(\lambda C^\nu - \phi L^\nu)(\lambda^* C^*_\nu - \phi^* L^*_\nu) = \frac{m^2}{4}(\lambda\lambda^*)(\phi\phi^*). \tag{2.103}$$

Since $\lambda\lambda^*$ and $\phi\phi^*$ are non-negative, we see that vector $U^\sigma$ becomes spacelike if $m^2 = -(\kappa_0)^2$ and that, from the second equation in (2.92),

$$\parallel \Omega \parallel = 8(\kappa_0)^2 \parallel V \parallel, \tag{2.104}$$

which shows that the scalar curvature associated with a spacelike CD flow field is negative. Using (2.96), we can show that (2.101) derived in a lightlike case remains exactly the same for spacelike cases.

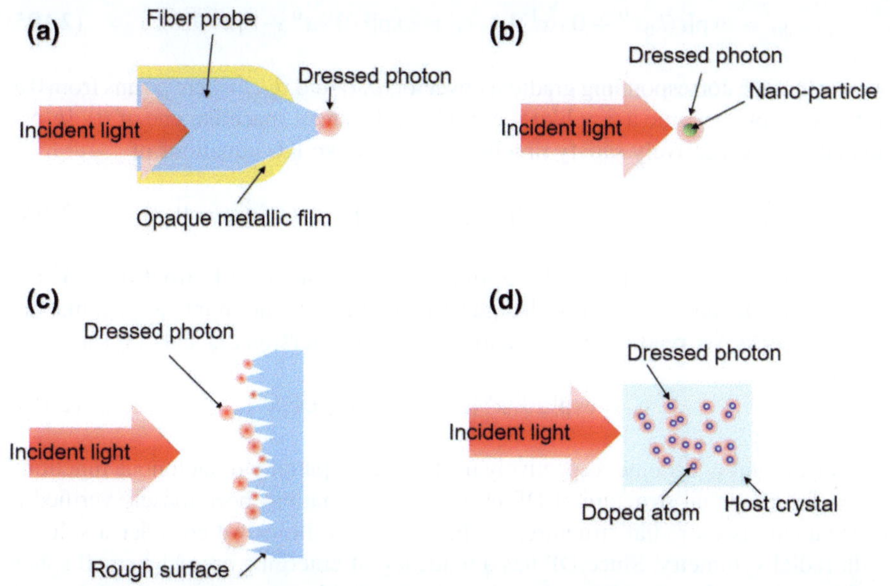


**Fig. 2.2** Typical experimental situations in which dressed photons emerge

## 2.5 New Model of Dressed Photon

Now, we are going to look into an important question of how the theoretical model discussed so far can actually be applied to an elusive picture of DP. Experimentally speaking, as is shown in Fig. 2.2, there are four typical situations in which DP emerges: (a) on a tip of a fiber probe, (b) on a nano-particle, (c) on a rough surface, (d) around doped atoms embedded in host crystal. Since case (b) seems to be the simplest one, let us consider DP field around such a given nano-particle. In doing so, it is important to keep in mind that DP field is not the manifestation of a certain on-shell field like streaming photon field. As its name indicates, it must be an "amalgamated" field arising from the interaction between incident light and a nano-particle's energy-momentum fields. The main reason why we consider spacelike momenta would be a key factor in understanding DP dynamics is simply because their existence is necessary for producing an extremely localized structure and a basic result of axiomatic quantum field theory [13] shows that field interactions generally accompany such spacelike 4-momenta, though, as Aharanov et al. [14] showed, they possess a noteworthy characteristic that a wavelike spacelike 4-momentum field becomes unstable when it encounters a timelike 4-momentum field.

Mathematically, the essence of this instability depends on the following well-known property of an analytic function $\exp[i(k_0 x^0 + k_1 x^1)]$. For a couple of functions defined below,

$$\varsigma_t \equiv \exp[i(k_0 x^0 + 0 \cdot x^1)], \quad \varsigma_x \equiv \exp[i(0 \cdot x^0 + k_1 x^1)], \tag{2.105}$$

we see that the corresponding gradient covector, $\partial_\mu \varsigma_t$ and $\partial_\mu \varsigma_x$ where $\mu$ runs from 0 to 1, respectively become a timelike vector $(ik_0\varsigma_t, 0)$ and a spacelike one of $(0, ik_1\varsigma_x)$, so that they respectively satisfy timelike and spacelike KG equations of

$$(\partial^2_{00} - \partial^2_{11})\varsigma_t + (k_0)^2\varsigma_t = 0, \quad (\partial^2_{00} - \partial^2_{11})\varsigma_x - (k_1)^2\varsigma_x = 0. \tag{2.106}$$

Now, let us see what happens if we *formally* invert the signs in front of $(k_0)^2$ and $(k_1)^2$, that is to say, we forcibly change a timelike KG equation into a spacelike one and vice versa. We readily see that, under such sign inversions, $\varsigma_t$ and $\varsigma_x$ respectively become

$$\varsigma_t = \exp[\pm k_0 x^0], \quad \varsigma_x = \exp[\pm k_1 x^1]. \tag{2.107}$$

Namely, $\varsigma_t$ and $\varsigma_x$ become respectively unstable and spatially monotonous functions.

As far as spatial structure of DP is concerned, what has been loosely verified in experiments is its radial structure, so that in what follows we consider a solution with radial symmetry. Since DP has a tendency of emerging around a small spherical boundary where medium changes abruptly, fields interaction at this boundary accompanying the bidirectional transformation between timelike and spacelike 4-momentum of the involved fields would also be abrupt, in which the mechanism similar to sign inversion between (2.105) and (2.107) is expected to be working. The assumption of radial symmetry reduces our DP model to a two-dimensional (2d) one in which we have $x^0, r$ as temporal and spatial coordinates. Notice that CP vortex field for which bivector $S_{\mu\nu}$ is expressed as a simple bivector of the form (2.31) is favorable to model the target dynamics since a couple of CP variables $\lambda$ and $\phi$ can respectively take care of the spatial and temporal structures of the interacting field. Employing a well-known separation of variable method, a general form of KG equation for $\psi = T(x^0)R(r)\Theta(\theta)\Phi(\varphi)$ where we assume $T(x^0) = \exp(ik_0x^0)$ and $\Phi(\varphi) = \exp(im\varphi)$ becomes

$$-(k_0)^2 - \left(\frac{R''}{R} + \frac{2}{r}\frac{R'}{R}\right) - \frac{1}{r^2}\left(\frac{\Theta''}{\Theta} + cot\theta\frac{\Theta'}{\Theta} - \frac{m^2}{sin^2\theta}\right) \pm (\kappa_\psi)^2 = 0, \tag{2.108}$$

where $\kappa_\psi$ is a real constant satisfying $\partial^\nu\partial_\nu\psi \pm (\kappa_\psi)^2\psi = 0$. Spherical symmetry of the solution means that $\Theta = const.$ and $m = 0$, so that we have

$$R'' + \frac{2}{r}R' + [(k_0)^2 \mp (\kappa_\psi)^2]R = 0. \tag{2.109}$$

Regarding $\lambda = R(r)$ and $\phi = T(x^0)$, a timelike KG equation for a spacelike vector $(0, R')$ and a spacelike KG equation for a timelike vector $(T', 0)$ respectively become

$$\text{timelike KG equation}: \quad k_0 = 0, \quad \Longrightarrow \quad R'' + \frac{2}{r}R' - (\kappa_r)^2 R = 0, \tag{2.110}$$

$$\text{spacelike KG equation}: \quad R = const., \quad \Longrightarrow \quad T'' - (\kappa_t)^2 T = 0. \tag{2.111}$$

Since our $2d$ DP problem have a couple of external parameters $d$ and $\omega$ which are the radius of a given nano-particle and the frequency of incident light, we specify $\kappa_r$ and $\kappa_t$ in the above system of equations by

$$\kappa_r = \frac{1}{d}, \quad \kappa_t = \frac{\omega}{c}, \quad \text{where } c \text{ denotes light velocity.} \tag{2.112}$$

It is important to note that although the orthogonality between $(0, R')$ and $(T', 0)$ corresponds to (2.42), a CP bivector field to be constructed from the above (2.110) and (2.111) is different from CD field defined in (2.101). This is because the form of energy-momentum tensor (2.99) satisfying a conservation law cannot be derived for the former due to the fact that CD field must be constructed from a couple of space like KG equations (2.86) satisfying

$$(m_\lambda)^2 - (m_\phi)^2 = 0 \tag{2.113}$$

which is explained in (2.91) and the following sentence.

As a short preparation for the physical interpretation of our solution, let us review the well-known formulation of Hamiltonian operator for a quantized harmonic oscillator. For the classical Hamiltonian of the form

$$H = \frac{p^2}{2m} + \frac{1}{2}m\omega^2 x^2, \tag{2.114}$$

where notations are conventional, the basis vectors of the stationary state energy representation $\mid n\rangle$ satisfy

$$\hat{H} \mid n\rangle = e_n \mid n\rangle, \tag{2.115}$$

where $e_n$ denote energy eigenvalues. It is advantageous to exchange the position $(\hat{x})$ and momentum $(\hat{p})$ operators for $\hat{a}$ and $\hat{a}^\dagger$ defined by

$$\hat{a} = \left(\frac{\omega m}{2\hbar}\right)^{1/2}\left[\hat{x} + \left(\frac{1}{\omega m}\right) i\hat{p}\right], \quad \hat{a}^\dagger = \left(\frac{\omega m}{2\hbar}\right)^{1/2}\left[\hat{x} - \left(\frac{1}{\omega m}\right) i\hat{p}\right], \tag{2.116}$$

which are respectively called energy lowering (annihilation) and raising (creation) operators. Using these operators, the Hamiltonian operator can be rewritten as

$$\hat{H} = \left(\hat{a}^\dagger\hat{a} + \frac{1}{2}\right)\hbar\omega, \qquad \epsilon_n = \left(n + \frac{1}{2}\right)\hbar\omega. \tag{2.117}$$

A particularly useful operator is $\hat{n} = \hat{a}^\dagger \hat{a}$ called the number operator that has the property

$$\hat{n}|n\rangle = \hat{a}^\dagger \hat{a}|n\rangle = n|n\rangle, \tag{2.118}$$

where $n$ denotes the number of quanta $\hbar\omega$ in the state.

Now, we go back to (2.110) and (2.111), and find the solutions to them:

$$\lambda = R(r) = \frac{1}{r}\exp\left(-\frac{r}{d}\right); \quad (0, R'), \tag{2.119}$$

$$\phi = \exp\left[\pm\left(\frac{\omega}{c}\right)x^0\right]; \quad (\phi', 0). \tag{2.120}$$

Equation (2.119) is known as Yukawa potential whose magnitude rapidly falls off as $r$ increases, which clearly corresponds to $\varsigma_x = \exp[-k_1 x^1]$ in (2.107), while (2.120) is exactly the same as $\varsigma_t = \exp[\pm k_0 x^0]$. Non-zero component of CP bivector field derived from (2.119) and (2.120) is $S_{01} = S_{0r}$ which becomes

$$S^\dagger_{0r} = \frac{\omega}{c}R'\exp\left[\frac{\omega}{c}x^0\right], \quad S_{0r} = -\frac{\omega}{c}R'\exp\left[-\frac{\omega}{c}x^0\right], \tag{2.121}$$

where $R' = -\frac{1}{d}\left(1 + \frac{d}{r}\right)R \neq 0$ and $-S^\dagger_{0r}S_{0r} = \left(\frac{\omega}{c}\right)^2(R')^2$. Since $S^\dagger_{0r}$ and $S_{0r}$ are monotonously increasing and decreasing functions of time $x^0$, qualitatively, the following analogy seems to be of help to understand the dynamics under consideration.

$$\hat{a}^\dagger|n\rangle = (n+1)^{1/2}|n+1\rangle \quad \Longleftrightarrow \quad S^\dagger_{0r} = \frac{\omega}{c}R'\exp\left[\frac{\omega}{c}x^0\right], \tag{2.122}$$

$$\hat{a}|n\rangle = (n)^{1/2}|n-1\rangle \quad \Longleftrightarrow \quad S_{0r} = -\frac{\omega}{c}R'\exp\left[-\frac{\omega}{c}x^0\right]. \tag{2.123}$$

Directly from the above equations, we see that $-S^\dagger_{0r}S_{0r}$ corresponds to the stationary state described by

$$\hat{n}|n\rangle = \hat{a}^\dagger\hat{a}|n\rangle = n|n\rangle \quad \Longleftrightarrow \quad -S^\dagger_{0r}S_{0r} = \left(\frac{\omega}{c}\right)^2(R')^2. \tag{2.124}$$

In order to understand a distinguished characteristic of and important role played by CD bivector field in the above discussion, we go back to the key condition (2.113). In Sect. 2.4.2, we discussed non-lightlike cases of CD field by assuming that a couple of Clebsch variables $\lambda$ and $\phi$ satisfy the same spacelike KG equation: $\partial^\nu\partial_\nu\psi - (\kappa_0)^2\psi = 0$. What has not been discussed there is a special case in which we have

$$\partial^\nu\partial_\nu\lambda + (\kappa_0)^2\lambda = 0, \quad \partial^\nu\partial_\nu\phi - (\kappa_0)^2\phi = 0. \tag{2.125}$$

As has been mentioned already, the reason why we exclude (2.125) is because it cannot be "on-shell"CD field satisfying a conservation law. An important characterisitcs of (2.125) is, however, that the associated $U^{\nu}U_{\nu}$ vanishes, which can be readily seen from (2.90). Namely, (2.125) may be regarded as an off-shell "light-like"mode of CP skew-symmetric field (2.30). In addition, a couple of exponentially amplifying and damping modes given in (2.121) that can be interpreted as off-shell "creation"and "annihilation"processes of virtual photons are also expressed in terms of CP representation. Thus we can say that the advantage of employing CP flows in understanding DP dynamics is twofold.

Firstly, it covers "on-shell"CD modes as a subset of it. Unlike a conventional transverse electromagnetic mode, a lightlike CD mode contains spacelike element of the form: $\partial^{\nu}\partial_{\nu}\lambda - (\kappa_0)^2\lambda = 0$, which means that, though by itself it is a free on-shell mode, it becomes unstable when it interacts with timelike 4-momentum and hence its lifetime would be extremely short. Recall that we also showed that a lightlike CD mode satisfies (2.72): the properties isomorphic to those of Riemann curvature tensor, which implies that a lightlike CD modes is an eigenmode possessing the property of the dynamical basis of $4d$ spacetime that can be excited as a sort of "resonant"mode in the sense of the most probable mode from the view point of Feynman's path-integral interpretation leading to the principle of the least action. We believe that all of those properties of lightlike CD modes are consistent to the behaviors of virtual photons.

Secondly, as is shown in the discussion concerning (2.125), CP flow field can provide a dynamical stage with off-shell character on which a couple of timelike and spacelike 4-momentum fields satisfying $\partial^{\nu}\partial_{\nu}\psi \pm (\kappa_0)^2\psi = 0$ merges into a null vector state $U^{\nu}U_{\nu} = 0$ that can be compared to a light field created by a pair annihilation. The existence of such null states together with the aforementioned properties of "creation"and "annihilation"of extended virtual photons is an inevitable ingredient of field interactions with off-shell character. Employing a newly introduced CP model, we can envisage a picture of DP dynamics as given incident light and nano-particle's field interactions in which (extended) virtual photons work as mediators adjusting the magnitudes of timelike and spacelike 4-momenta involved in the interactions.

## 2.6 Concluding Remarks and on Extended Topics

Based on the unified view of MMDS [7] connecting a conventional interpretation of covariant quantization of electromagnetic wave fields and the associated scalar and longitudinal modes, we have elaborated a novel classical model of virtual photons which play a key role in electromagnetic Coulomb interactions. In order to embody the model, we utilized the concept of CP flow with a couple of scalar parameters $\lambda$ and $\phi$ originally introduced in the study of Hamiltonian structure of barotropic fluid motions and applied it to electromagnetic 4-vector potential $A^{\mu}$ of free propagating modes. It turns out that CP bivector field defined as the curl of CP flow yields an intriguing mathematical structure which is quite similar to that of free electromagnetic wave field $F^{\mu\nu}$ if we impose certain auxiliary conditions on $\lambda$ and $\phi$. We call CP bivector field with those constraints CD (free electromagnetic wave) field, since it satisfies $\partial^{\nu}\partial_{\nu}U^{\sigma} - (\kappa_0)^2 U^{\sigma} = 0$: a spacelike KG equation which can be regarded as a dual of Proca equation $\partial^{\nu}\partial_{\nu}A^{\sigma} + (\kappa_0)^2 A^{\sigma} = 0$ with respect to the sign inversion of $(\kappa_0)^2$. In parallel with this line of research, we also have explored the possibility that, unlike a conventional interpretation, $\partial_{\nu}A^{\nu}$ is a physical quantity, and showed that, for a lightlike case of CD field, such a possibility can be realized without violating the energy-momentum conservation law. If we identify one of Clebsch variables $\phi$ as $\partial_{\nu}A^{\nu}$, then CD field originally defined as a special class of vortex field associated with the congruence of null geodesics becomes a dual field of electromagnetic free wave in the sense explained above accompanying a longitudinal mode.

A unique feature of CD field is that, in addition to $\phi \equiv \partial_{\nu}A^{\nu}$, $\lambda$ satisfies a spacelike KG equation (2.36). Such a spacelike KG equation seems to be quite foreign to a conventional concept of free electromagnetic wave. In Sect. 2.5, we referred to the instability of spacelike modes put in an environmental timelike 4-momentum field. Considering this fact into consideration, however, we think that the instability of CD field in such an environment is a rather preferable feature for us to identify CD field as a model of virtual photons which are characterized as photons with extreme ephemerality and with longitudinal electric field. In this regard, based on his analyses of electromagnetic response theory, Banno [15] emphasized that non-resonant longitudinal electric components are inevitable ingredients for the emergence of optical near field and DP, which seems to be consistent with our picture of DP modeled by CD field.

In Sect. 2.4.2, we have referred to the extension of lightlike CD field to the case of spacelike one and showed that its vortical structure is isomorphic to that of Riemann curvature tensor, which means that, aside from its instability arising from the interaction with timelike 4-momenta, by itself, it is an on-shell mode satisfying the conservation of energy-momentum and is also a dynamical base state of spacetime as in the case of a plane wave in Fourier space. An intriguing aspect of spacelike CD field is that, corresponding to $U^{\nu}U_{\nu} < 0$, its accompanying scalar curvature is negative which means a negative energy state. Contrary to a prevailing blind belief on the one-sided energy spectrum of state vectors, a justification of such a negative energy state is given in the final section of SOO1 by referring to a modular inversion symmetry in Tomita-Takesaki extension [16] of thermal equilibrium to infinite dimensional systems. Notice that, from the view point of Ojima's work [6] on the interpretation of longitudinal mode of electromagnetic wave, $\lambda$ of CD field satisfying spacelike KG equation in a lightlike case is an entirely new ingredient that was not touched on in his arguments, which implies a possibility of $\lambda$ being quantized. When we go back to (2.49) and appreciate the dual expression of $\hat{T}^{\nu}_{\mu}$, especially the second one of $\rho C_{\mu}C^{\nu}$, this conjecture seems to be of special interest since $\hat{T}^{\nu}_{\mu}$ is isomorphic to the energy-momentum tensor of free moving (fluid) particles for which density $\rho$ is discretized for a real fluid. If this is a case, then, there is a high possibility that the quantized $\lambda$ field satisfies, for instance, spacelike Dirac's equation and the spinor field associated with it behaves as a fermion which may be interpreted as the dual entity of an electron in timelike domain. This picture seems to be a new version of Dirac's sea idea which may give a clue to understand the enigmatic dark energy discussed in cosmology. A preliminary survey on this issue will be given elsewhere.

**Acknowledgements** The author is thankful to all the members who have participated in nonregular internal workshops of RODreP for their inspiring discussions and special thanks to Dr. Ohtsu for his distinguished leadership for dressed photon study project and to Dr. Ojima for his valuable theoretical guidances in varieties of research subjects. A part of this work was supported by Research Foundation for Opto-Science and Technology.

## References

1. H. Sakuma, I. Ojima, M. Ohtsu, PROG Q ELEC **55**, 74–87 (2017)
2. H. Sakuma, I. Ojima, M. Ohtsu, Appl. Phys. A **123**, 750 (2017). https://doi.org/10.1007/s00339-017-1364-9
3. M. Ohtsu, K. Kobayashi, *Optical Near Fields* (Springer, Heidelberg, 2004)
4. M. Ohtsu, *Dressed Photons* (Springer, Berlin, 2014)
5. Chapter 2, in *Dressed Photons*, (Springer, Berlin, 2014), pp. 11–36
6. I. Ojima, Nananishi-Lautrup B field, crossed product and duality, in *Research on Quantum Field Theory, RIMS Workshop, Kyoto, 2006*, pp. 29–37
7. I. Ojima, Micro-macro duality in quantum physics, in *Stochastic Analysis: Classical and Quantum Perspective of White Noise Theory*. ed. by T. Hida (World Scientific, Singapore, 2005), pp. 143–161
8. N. Nakanishi, Prog. Theor. Phys. **35**, 1111 (1966)
9. A. Tonomura, N. Osakabe, T. Matsuda, T. Kawasaki, J. Endo, S. Yano, H. Yamada, H. Phys, Rev. Lett. **56**, 792–795 (1986)
10. S.H. Lamb, *Hydrodynamics*, 6th edn. (Cambridge University Press, Cambridge, 1930)
11. I. Ojima, *In Private Communications*
12. S.A. Huggett, K.P. Tod, *An Introduction to Twistor Theory*, 2nd edn. (Cambridge University Press, Cambridge, 1994)
13. R. Jost, *The generalized theory of quantized fields,* (Amer. Math. Soc. Publ. Providence, 1963)
14. Y. Aharonov, A. Komar, L. Susskind, Phys. Rev. **182**, 1400–1403 (1969)
15. I. Banno, M. Ohtsu, Theory II on non-linear resonance effect in optical near field: attempt on understanding the dynamics of dressed photon by response theory, in *78th JSAP Autum Meeting at Fukuoka*, Japan, September 5–8, 2017, Oral Presentation Number: 7a-A405-9 (2017)
16. O. Bratteli, D. Robinson, *Operator Algebras and Statistical Mechanics*, vol. 1, 2nd edn. (Springer, Berlin, 1987)

# Chapter 3
# Quantum Probability for Dressed Photons: The Arcsine Law in Nanophotonics

**Hayato Saigo**

**Abstract** In the present paper we propose a new framework for the investigation of dressed photon from the viewpoint of quantum probability. As a result, we show that a probability measure called the Arcsine law plays crucial roles in nanophotonics. For instance, a concrete formula for the probability density of dressed photon in a fiber probe and a new prediction in photon breeding phenomena are given in terms of the Arcsine law.

## 3.1 Introduction

*Dressed Photon* (DP) is an essential concept for the investigation of many phenomena which cannot be explained in conventional optics, especially of the phenomena arising in the interaction between light and nano-scale matter [1]. This concept provides the fundamental reflections on physics and engineering and the interesting examples motivating the reformulation of the concept such as the composite systems or the micro-macro relationships.

In this paper we investigate some mathematical/physical aspects of the DP in terms of *Quantum Probability* (QP), the generalization of probability theory which includes whole quantum (and classical) physics as a subtheory.

As a result, it is shown that an important probability distribution called the *Arcsine law* plays fundamental roles in the phenomena of DP, and a new mathematical/physical interpretation is given for the universal phenomenon that DP occurs at the extreme points of a material system such as the tip of a fiber probe.

In the Sect. 3.2 we define the notion of DP from the author's viewpoint which clarifies why DP is related to the fundamental notions of physics such as *quantum-*

H. Saigo (✉)
Nagahama Institute of Bio-Science and Technology, Nagahama, Japan
e-mail: harmoniahayato@gmail.com



T. Yatsui (ed.), *Progress in Nanophotonics 5*, Nano-Optics and Nanophotonics,
https://doi.org/10.1007/978-3-319-98267-0_3

*classical correspondence*. The following Sect. 3.3 is devoted to discuss conceptual issues which lead to the concepts of QP.

In the Sect. 3.4 we introduce basic notions in QP such as *quantum probability space* in mathematical language. The notion of *algebraic random variables* and their *moments*, which connects QP to probability measures and statistics in terms of conventional probability theory, are introduced in the Sect. 3.5. In the following Sect. 3.6, the *GNS representation*, which bridges the notions in QP and the ones in conventional quantum physics based on Hilbert spaces, is introduced. These three chapters about basic concepts in QP is mainly based on [2]. (For more precise introduction, see [2] and the references therein.)

The Sects. 3.7 and 3.8 treats the main topic in the present paper; *quantum-classical correspondence*. Our main result is the ubiquity of the Arcsine Law in quantum physics, in the context of micro-macro relationships. The contents of these chapters are discovered and discussed in [3, 4].

In the Sect. 3.9 we apply the result to the study of of DP. It is shown that the probability distribution of DP will be the densest around the extreme points of the matter, from the results in Sects. 3.7 and 3.8. The problem of impurity in the matter is also mentioned, and the new theoretical prediction about DP is proposed.

The Sect. 3.10 provides the conclusion of the paper. In the appendices, we give a very short introduction to some more mathematical aspects of quantum probability and basic knowledge on moments in conventional probability. (The two appendices are also based on [2].)

## 3.2 What is Dressed Photon?

Dressed Photons are a new kind of photons localized in space-time and having width in momentum-energy. It is known that massless particles with finite non-zero spin, such as free photons, cannot be localized in space, in the meaning that even the position operators cannot be well-defined [5, 6]. When we take the effective mass of photons generated by the interaction between light and matter into account, however, it turns out to be natural to consider localized photons. Especially in the case that the materials are in the nano-scale, the localization in space-time and the fluctuation in momentum-energy provides brand new aspects of light. The photon in such context is called *Dressed Photons* (DP) [1, 7] (Figs. 3.1 and 3.2).

For the theoretical definition of DP, we need to consider the *off-shell* nature of interaction. As is well known, you cannot formulate quantum field theories without entities off the mass-shell. In other words, traditional particle description fails to hold in the treatment of the composite systems of quantum fields which interact each other. Hence, DP must be entities so different from the Einstein's quanta of lights, or free photons. For example, DP must have infinitely many modes.

Here a fundamental question arises: How can we understand the DP as an individual entity? As far as you stick to the notion of individuals as irreducible particles on the mass-shell, it is impossible to treat DP as individual. However, from more gen-

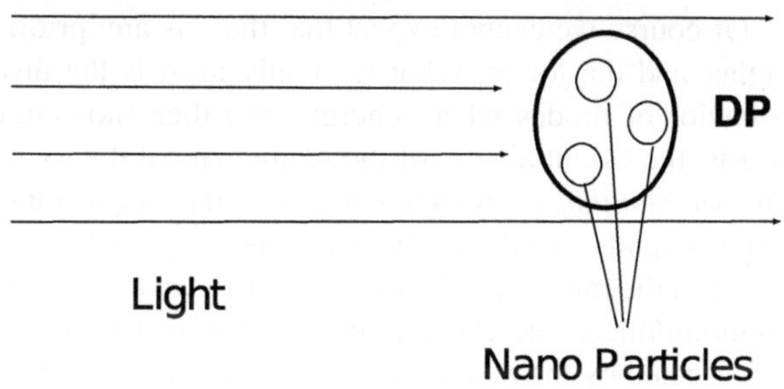


**Fig. 3.1** Intuitive description of DP

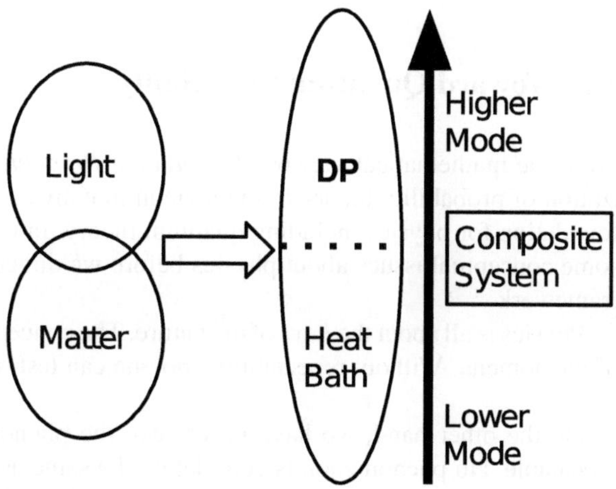


**Fig. 3.2** Our definition of DP

eral perspective advocated by Ojima [8], macroscopic physical phenomena can be emerged out of condensation of microscopic off-shell entities. In this line of thought we propose a basic idea:

> Basic Idea: In the interaction between light and (nano-scale) matter, certain families of modes of the composite system will become to behave like individuals.

Then another question arises: How should we identify certain families of modes as individuals? As a result of the upper- and the downer- conversion of modes due to the interaction between light and matter, the width of energy-momentum becomes larger for the composite system rather than that of originally injected light [1]. Then it is natural to focus the family of the higher energy-momentum modes and of the lower energy-momentum modes for the investigation of the phenomena which cannot be explained in conventional optics where everything is considered to be on the mass-shell. In the present paper, we propose to identify the family of higher modes *as* DP, for which the family of lower modes play a kind of heat-bath. In other words, we propose our definition of DP as a working hypothesis:

> Our definition of DP: In the interaction between light and (nano-scale) matter, the family of higher modes of the composite system become to behave like an individual. We call it the dressed photon field or simply the dressed photon (DP).

Of course we cannot expect that there is any priori strict boundary between the higher and the lower. What we really need is the investigation on the asymptotic behavior of modes where energy (and then momentum) becomes large. In other words, the essential core of the mathematical theory for DP theory is nothing but a kind of *quantum-classical correspondence*, i.e. asymptotics of states as the quantum number tends to infinity. In that sense, DP will appear in the intermediate realm between the micro and the macro. Hence we need some general framework for both quantum/micro and classical/macro. Fortunately, quite a suitable theory have been constructed by the mathematicians, which is called *Quantum Probability* (QP).

## 3.3 Toward Quantum Probability

From the mathematical viewpoint, *Quantum Probability* (QP ) is such a generalization of probability theory to the noncommutative context. It also gives a unified foundation for physics including quantum theory. In the present section we discuss some conceptual issues about physics before we directly get into the mathematical framework.

Physics is all about the *laws* of the nature. The concept of law implies repeatability of phenomena. Without repeatability, no one can test or state the correctness of the law.

On the other hand, we have never seen the phenomena which are completely repeatable. No phenomenon is completely the same as another phenomenon. What we really mean by the concept of law obtained through experiences is as follows:

> If similar phenomena arise, then we can predict or control the development of the phenomena to some extent.

The similarity is not a priorly given, nor unique. It depends on our criterion for *equivalence relation* between phenomena and tested by continual experiences. Moreover, in many cases, our prediction or control for the development are not, cannot be or should not be, deterministic. In short, our concept of law is essentially *statistical* one based on equivalence relation we construct between phenomena.

For the construction of the equivalence relation between phenomena, we make use of the notion of *system* and *situation*. The term situation is nothing but the relationship to an *environment*, which can be considered as another system around the system in focus (Fig. 3.3).

Even though all phenomena we experience are impermanent, we see "relatively constant" phenomena within them. More precisely, we see somekind of the sameness between phenomena at each moments, and then consider as if there is "something the same" having own properties. This "something the same" between the development of phenomena is called system. Then a lot of phenomena can be considered as one system in the different situation, or in the different relationships between the system in focus and the environment around it. Of course the sameness between situations is also depending on another equivalence relation, as environments are also considered

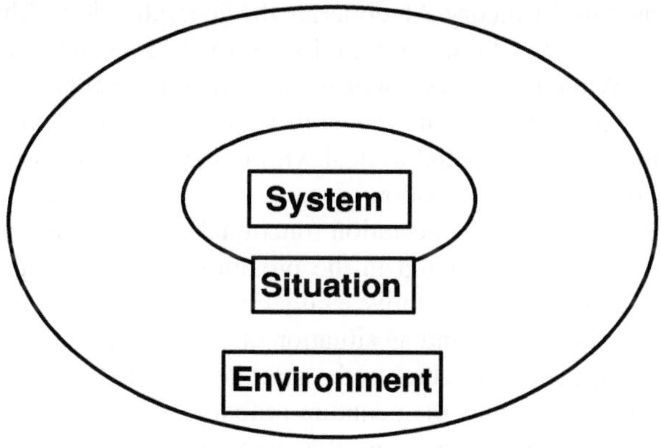


**Fig. 3.3** System and situation

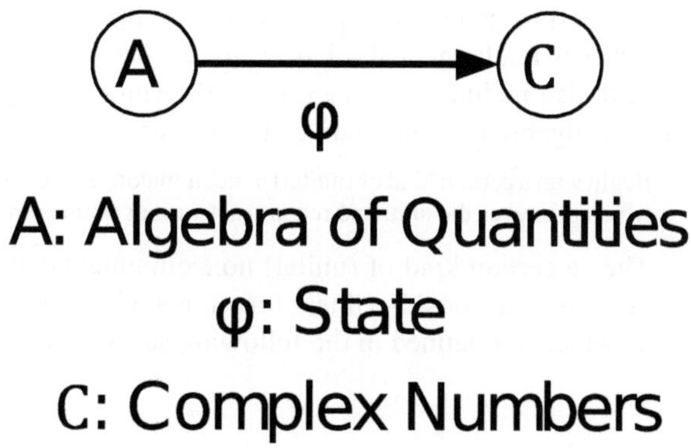


**Fig. 3.4** Concept of quantum probability space

as systems and the relationship between the two system can be different depending on the *interface* between them.[1]

The central concept of quantum probability is *quantum probability space* (Fig. 3.4). It is a pair of *algebra* and *state*, which are the mathematical counterpart of the concept of the system and the situation, or the system-environment interface. In the following section we introduce the mathematical definition of the quantum probability space.

## 3.4 Quantum Probability Space

### *3.4.1 What is Quantum Probability Space?*

In the present section we introduce the notion of algebras and states on them to define the notion of quantum probability space, which is the central notion in quantum

[1]For example, there are no completely the same experiment in completely the same environment. Nevertheless we assume some equivalence. The equivalence between the experimental environments are in general defined thorough the *protocol* of the experiments.

probability theory. Algebras are the generalization of the algebra of random variables in the conventional probability theory originated by Kolmogorov [9].

While in the Kolmogorovian probability theory the algebras of random variables as functions on some probability space are commuting with each other, in quantum theory it is essential to deal with noncommutative random variables which we call observables. The notion of states in modern physics, including quantum theory, are nothing but the expectation functionals for noncommutative random variables.

As we have noted in the previous section, an algebra (more concretely, a not necessarily commutative *-algebra defined below) and a state on it describes the system in focus and an situation or system-environment interface.[2] Then the pair of an algebra (*-algebra) and a state represent the statistical law of the system in the situation. That is the quantum probability space.

Apart from the name, the scope of the notion of quantum probability space is not limited to quantum physics. For example, it has many applications in various branches in mathematical sciences, for example, spectral analysis of growing graphs [10]. It also includes in principle all the contents of probability theory as the case that the algebra is commutative. The slogan of quantum probability is:

> Begin with a certain kind of (unital) noncommutative algebra and a positive linear functional which preserves the unit, and reconstruct various notions from them.

The "a certain kind of (unital) noncommutative algebra" is the nothing but a *-algebra, and the "positive linear functional which preserves the unit" nothing but the state, which are defined in the following subsections.

### 3.4.2 *-Algebras

**Definition 1** Let $\mathcal{A}$ be a vector space over $\mathbb{C}$. It is said to be an algebra (over $\mathbb{C}$) it it satisfies the following:

(i) it has a product $(a, b) \in \mathcal{A} \times A \mapsto ab \in \mathcal{A}$ which is distributive, i.e.,

$$a(b+c) = ab + ac, \quad (a+b)c = ac + bc, \quad \lambda(ab) = (\lambda a)b = a(\lambda b)$$

for all $a, b \in \mathcal{A}$ and $\lambda \in \mathbb{C}$.

(ii) the product is associative, i.e.,

$$(ab)c = a(bc)$$

for all $a, b, c \in \mathcal{A}$.

**Definition 2** An algebra over $\mathcal{A}$ is said to be unital if

(iii) the product has the unit $1_{\mathcal{A}}$.

---

[2]This view point has been advocated by Ojima [8] for a long time.

**Definition 3** An algebra $\mathcal{A}$ is said to be commutative if

$$ab = ba$$

holds for $a, b \in \mathcal{A}$. Otherwise, it is said to be noncommutative.

**Definition 4** A mapping $f : \mathcal{A} \longrightarrow \mathcal{B}$ between algebras (resp. unital algebras) is said to be a homomorphism if it satisfies (i)–(iii) (resp. (i)–(iv)) below:

(i) $f(a+b) = f(a) + f(b)$
(ii) $f(\lambda a) = \lambda f(a)$
(iii) $f(ab) = f(a)f(b)$
(iv) $f(1_{\mathcal{A}}) = 1_{\mathcal{B}}$

hold for all $a, b \in \mathcal{A}$ and $\lambda \in \mathbb{C}$.

If $f$ has an inverse mapping which is a homomorphism, it is said to be an isomorphism. If $f$ is injective, it is said to be a monomorphism. If there is an isomorphism between $\mathcal{A}$ and $\mathcal{B}$, then $\mathcal{A}$ and $\mathcal{B}$ is said to be isomorphic. If there is an monomorphism $\mathcal{A}_0$ to $\mathcal{A}$, then $\mathcal{A}_0$ (or the monomorphism itself) is said to be a subalgebra of $\mathcal{A}$.

**Definition 5** An algebra (resp. unital algebra) $\mathcal{A}$ is said to be a *-algebra (resp. unital *-algebra) if it has an involution operation $* : \mathcal{A} \longrightarrow \mathcal{A}$ which satisfies the following:

(i) $(a+b)^* = a^* + b^*$
(ii) $(\lambda a)^* = \bar{\lambda} a^*$
(iii) $(ab)^* = b^* a^*$
(iv) $(a^*)^* = a$

A homomorphism $f : \mathcal{A} \longrightarrow \mathcal{B}$ between (unital) *-algebras are said to be a *-homomorphism if $f(a^*) = f(a)^*$ holds for all $a \in \mathcal{A}$. If $f$ has the inverse *-homomorphism, it is said to be a *-isomorphism. If $f$ is injective, it is said to be a *-monomorphism. If there is a *-isomorphism between $\mathcal{A}$ and $\mathcal{B}$, then $\mathcal{A}$ and $\mathcal{B}$ is said to be *-isomorphic. If there is an *-monomorphism $\mathcal{A}_0$ to $\mathcal{A}$, then $\mathcal{A}_0$ is said to be a *-subalgebra of $A$.

It is easy to see that $1_{\mathcal{A}}^* = 1_{\mathcal{A}}$. $\mathcal{A} = \{0\}$ becomes a trivial *-algebra, called zero algebra. As for a (nonzero) unital *-algebra $\mathcal{A}$, $\lambda \in \mathbb{C} \mapsto \lambda 1_{\mathcal{A}} \in \mathcal{A}$ becomes a *-monomorphism. Then $\mathbb{C}$ is considered as a subalgebra of $\mathcal{A}$ and $\lambda 1_{\mathcal{A}}$ is simply denoted as $\lambda$.

### 3.4.3 State

**Definition 6** Let $\mathcal{A}$ be a unital *-algebra. A linear functional $\varphi : \mathcal{A} \longrightarrow \mathbb{C}$ is said to be a state on $\mathcal{A}$ if it satisfies the following:

(i) $\varphi(1) = 1$
(ii) $\varphi(a^*a) \geq 0$ for all $a \in \mathcal{A}$.

**Proposition 1** *Let $\varphi$ be a state on $\mathcal{A}$. Then*

$$\varphi(a^*) = \overline{\varphi(a)}$$

*holds for all $a \in \mathcal{A}$. In other words, $\varphi$ is a $^*$-morphism.*

### *3.4.4 Quantum Probability Space*

**Definition 7** A pair $(\mathcal{A}, \varphi)$ composed of a unital $^*$-algebra $\mathcal{A}$ and a state on it is called an quantum probability space. When $\mathcal{A}$ is commutative, $(\mathcal{A}, \varphi)$ is said to be classical probability space.

**Definition 8** Let $(\mathcal{A}, \varphi)$ and $(\mathcal{B}, \psi)$ be two quantum probability spaces. They are said to be isomorphic if there exists some $^*$-isomorphism $f : \mathcal{A} \longrightarrow \mathcal{B}$ which satisfies $\varphi = \psi \circ f$.

*Example 1* Let $\mathcal{H}$ be a Hilbert space and $\mathcal{B}(\mathcal{H})$ be the unital $^*$-algebra of all bounded linear operators on $H$. For a unit vector $\Omega \in \mathcal{H}$, we define a state $\varphi_\Omega$ by

$$\varphi_\Omega(a) := <\Omega, a\Omega>.$$

which is called a vector state associated to $\Omega$ and often denoted simply as $\Omega$.

Then $(\mathcal{B}(\mathcal{H}), \Omega)$ becomes an quantum probability space.

The modified versions of $(\mathcal{B}(\mathcal{H}), \Omega)$ will turn to be universal in Sect. 3.6.

*Example 2* Let $(\Omega, \mathcal{F}, P)$ be probability space in the meaning of conventional (Kolmogorovian, or measure-theoretic) probability theory. Then we define a classical probability space $(\mathcal{M}_\infty(\Omega), E)$ by

$$\mathcal{M}_\infty(\Omega) := \cap_{1 \leq p < \infty} L^p(\Omega)$$

which is naturally equipped with the involution (complex conjugate) and

$$E(X) := \int_\Omega X(\omega) P(d\omega)$$

which is well defined on $\mathcal{M}_\infty(\Omega)$. Note that any probabilistic notion on $(\Omega, \mathcal{F}, P)$ can be defined by the indicator functions, which are the members of $L^\infty(\Omega)$. Since $L^\infty(\Omega) \subset \mathcal{M}_\infty(\Omega)$, an algebraic probability space $(\mathcal{M}_\infty(\Omega), E)$ contains all the essential information of $(\Omega, \mathcal{F}, P)$.

*Example 3* Let $(\Omega, \mathcal{F}, P)$ be a probability space in the meaning of conventional probability theory and $\mathbb{C}[X]$ be the $^*$-algebra generated by $X \in \mathcal{M}_\infty(\Omega)$. Then we call $(\mathbb{C}[X], E)$ the quantum probability space generated by $X$, which is obviously classical probability space. This classical probability space is useful to analyze random variables in conventional probability theory or probability measures from the noncommutative viewpoint, as is discussed in Sect. 3.6.

The example above suggests that we can "recover" conventional probability space from classical probability space in the meaning of quantum probability theory. In fact, for the case that the $^*$-algebra is an $C^*$-*algebra*, from a classical probability space we can obtain a conventional measure-theoretic probability space. (For the precise statement, see the Appendix A). In short, quantum probability theory includes the conventional probability theory as a quite meaningful subsystem.

## 3.5 Algebraic Random Variables and Moments

### *3.5.1 Algebraic Random Variables*

**Definition 9** Let $(\mathcal{A}, \varphi)$ be an quantum probability space. An element $a \in \mathcal{A}$ is said to be an algebraic random variable in $(\mathcal{A}, \varphi)$. If $a = a^*$, it is said to be real.

*Remark 1* In some contexts, an algebraic random variable in $(\mathcal{A}, \varphi)$ is defined as a $^*$-morphism from some unital $^*$-algebra $\mathcal{A}_0$ to $\mathcal{A}$. This generalized definition is considered to have some advantages, especially in dealing random variable which does not admit finite moments. For the discussions in the present paper, however, the definition we adopt above is sufficient. (For details, see [11], chapter 1.)

### *3.5.2 Moments*

**Definition 10** Let $a$ be an algebraic random variable. Quantities such as

$$\varphi(a^{\epsilon_1} \cdots a^{\epsilon_m}), \quad \epsilon_i \in \{1, *\}$$

are called mixed moments of $a$. For a real algebraic random variable $a$, any mixed moment is reduced to the form

$$\varphi(a^m)$$

which is called the $m$th moment moment of $a$ and denoted as $M_m(a)$.

As statistical information of each algebraic random variable as themselves (but not of the relations with other algebraic random variables) is condensed in the moments, it is natural to introduce an equivalence relation between algebraic random variables as follows.

**Definition 11** Let $(\mathcal{A}, \varphi)$ and $(\mathcal{B}, \psi)$ be algebraic probability spaces. An algebraic random variable $a$ in $(\mathcal{A}, \varphi)$ and another algebraic $b$ random variable in $(\mathcal{B}, \psi)$ are said to be stochastically equivalent if

$$\varphi(a^{\epsilon_1} \cdots a^{\epsilon_m}) = \psi(b^{\epsilon_1} \cdots b^{\epsilon_m})$$

holds for all $m \in \mathbb{N}$ and $\epsilon \in \{1, *\}^{\underline{m}}$. If $a, b$ are real algebraic random variables, it simply means that

$$M_m(a) = M_m(b)$$

holds for all $m \in \mathbb{N}$.

### 3.5.3 Relation to Probability Measures

For real random variables, the following theorem holds:

**Theorem 1** *Let $(\mathcal{A}, \varphi)$ be an quantum probability space and a be a real algebraic random variable in it. Then there exists a probability measure on $\mathbb{R}$ such that*

$$M_m(a) := \varphi(a^m) = \int_{\mathbb{R}} x^m \mu(dx)$$

*for all $m \in \mathbb{N}$.*

The quantity $\int_{\mathbb{R}} x^m \mu(dx)$ is called moments of the random variable $x$ obeying the probability measure $\mu$, also in conventional probability or statistics (Some basic knowledge on moments known in conventional probability theory and statistics are given in Appendix "Moments in Probability Theory".).

This terminology is coherent with our definition of moments of algebraic random variables: All random variables (with all finite moments) in conventional probability theory can be considered as algebraic random variables, and the two notions of moments are equivalent for conventional (measure-theoretic) notion of random variables. The meaning of the theorem above is, is short, any real algebraic random variable is stochastically equivalent to some measure-theoretical random variable on $R$ in the meaning of conventional probability theory.

Here we adopt an important notation for a state $\varphi : \mathcal{A} \to \mathbb{C}$, an element $X \in \mathcal{A}$ and a probability distribution $\mu$ on $\mathbb{R}$.

**Notation**: $X \sim_{\varphi} \mu \iff \varphi(X^m) = \int_{\mathbb{R}} x^m \mu(dx)$ for all $m \in \mathbb{N}$.

The essential point in the proof of Theorem 1 is that we can deduce the existence of $\mu$ from the positivity of $\varphi$, by the use of Hamburger's theorem on the characterization of moment sequences. (see [10], for details.)

*Remark 2* The uniqueness of such probability measure $\mu$, however, does not hold in general. (The determinate moment problem discussed in the Appendix B.)

**Definition 12** $\mu$ above is called a distribution of $a$. (If the uniqueness holds, we should call it *the* distribution of $a$.)

Hence, as for *one* real algebraic random variable, notions of conventional, or measure-theoretic, probability is available. This fact is a typical and elementary example of the passage *from quantum to classical.*

### 3.5.4 Uncertaninty Principle

However, the "non-classicality" appears when we proceed to the discussion on two (or more) algebraic random variables which are not commutative with each other. As a typical and important example, let us discuss the uncertainty relation between the variance of two noncommuting observables, taken as the most famous feature of quantum theory.

The mathematical essence of the uncertainty relation in the above sense is a consequence of a version of Schwarz inequality below:

**Theorem 2** *Let $(\mathcal{A}, \varphi)$ be an quantum probability space. Then*

$$|\varphi(a^*b)|^2 \leq \varphi(a^*a)\varphi(b^*b)$$

*for all $a, b \in \mathcal{A}$.*

**Definition 13** Let $a$ be an algebraic random variable in $(\mathcal{A}, \varphi)$. The variance $\Delta_\varphi(a)$ of $a$ is defined as follows:

$$\Delta_\varphi(a)^2 = \varphi((a - \varphi(a))^*(a - \varphi(a))) = \varphi(a^*a) - |\varphi(a)|^2$$

**Lemma 1** *For all algebraic random variable $a, b$ in $(\mathcal{A}, \varphi)$*

$$|\varphi([a, b])| \leq \Delta_\varphi(a^*)\Delta_\varphi(b) + \Delta_\varphi(a)\Delta_\varphi(b^*).$$

*(Here $[a, b]$ denotes the commutator $ab - ba$.)*

*Proof* Note that $[a - \varphi(a), b - \varphi(b)] = [a, b]$ and $\Delta_\varphi(a - \varphi(a)) = \Delta_\varphi(a)$. Then the lemma directly follows from Theorem 2.

As a consequence we obtain the uncertainty relation:

**Theorem 3**

$$|\varphi([a, b])| \leq 2\Delta_\varphi(a)\Delta_\varphi(b)$$

*holds for all real algebraic random variables $a, b$ in $(\mathcal{A}, \varphi)$.*

## 3.6 Representation of Quantum Probability Space

### 3.6.1 Representation of *-Algebra

**Definition 14** Let $\mathcal{H}$ be a Hilbert space and $\mathcal{D}$ be a dense subspace of $\mathcal{H}$. Then we define a unital *-algebra $\mathcal{L}(\mathcal{D})$ associated to $\mathcal{D}$ as

$$\mathcal{L}(\mathcal{D}) := \left\{T : \mathcal{D} \longrightarrow \mathcal{D}, linear;\ \mathcal{D} \subset Dom(T^\dagger) \text{ } and \text{ } T^\dagger(\mathcal{D}) \subset \mathcal{D}\right\} \tag{3.1}$$

with involution $T^* := T^\dagger|_{\mathcal{D}}$. (Here $^\dagger$ denotes the conjugate operation and $Dom(A)$ denotes the domain of $A$.)

**Definition 15** Let $\mathcal{A}$ be a unital *-algebra, $\mathcal{H}$ be a Hilbert space and $\pi : \mathcal{A} \longrightarrow \mathcal{L}(\mathcal{D})$ be a unital *-homomorphism. Then the triple $(\mathcal{H}, \mathcal{D}, \pi)$ is said to be a representation of $\mathcal{A}$ on a Hilbert space.

**Definition 16** Let $\mathcal{A}$ be a unital *-algebra and $(\mathcal{H}, \mathcal{D}, \pi)$ be a representation of $\mathcal{A}$ on a Hilbert space. A vector $\Phi \in \mathcal{D}$ is called a cyclic vector if $\pi(\mathcal{A})\Phi := \{\pi(a)\Phi; a \in \mathcal{A}\}$ is dense subspace of $\mathcal{H}$. When $\Phi$ is a cyclic vector, a quadruple $(\mathcal{H}, \mathcal{D}, \pi, \Phi)$ is said to be a cyclic representation of $\mathcal{A}$.

**Definition 17** Two cyclic representation $(\mathcal{H}_i, \mathcal{D}_i, \pi_i, \Phi_i)$, $i = 1, 2$, of $\mathcal{A}$ said to be unitary equivalent if there exist some unitary operator $U : \mathcal{H}_1 \longrightarrow \mathcal{H}_2$ such that

$$U\pi_1(a)\Phi_1 = \pi_2(a)\Phi_2 \tag{3.2}$$

holds for all $a \in \mathcal{A}$.

**Proposition 2** *If two cyclic representation $(\mathcal{H}_i, \mathcal{D}_i, \pi_i, \Phi_i)$, $i = 1, 2$, of $\mathcal{A}$ satisfies*

$$\langle\Phi_1, \pi_1(a)\Phi_1\rangle = \langle\Phi_2, \pi_2(a)\Phi_2\rangle \tag{3.3}$$

*for all $a \in \mathcal{A}$, then they are unitary equivalent.*

### 3.6.2 GNS Representation

**Theorem 4** *For any quantum probability space $(\mathcal{A}, \varphi)$, there exists a cyclic representation $(\mathcal{H}, \mathcal{D}, \pi, \Phi)$ of *-algebra $\mathcal{A}$ which satisfies*

$$\varphi(a) = \langle\Phi, \pi(a)\Phi\rangle \tag{3.4}$$

*for any $a \in \mathcal{A}$. Such representation is unique up to unitary equivalence.*

**Definition 18** $(\pi(\mathcal{A}), \Phi)$ above becomes an algebraic probability space. We call it a GNS (Gel'fand–Naimark–Segal) representation of $(\mathcal{A}, \varphi)$.

*Remark 3* If $\varphi$ is faithful (i.e., $\varphi(a^*a) = 0$ implies $a = 0$), $(\mathcal{A}, \varphi)$ and its GNS representation $(\pi(\mathcal{A}), \Phi)$ is isomorphic.

The notion of GNS representation has vast implications. First of all, it connects conventional formulation of quantum theory based on Hilbert space with quantum probability theory, which works also as a algebraic framework for quantum theory.

Moreover, it also shed new light on the conventional probability theory. Given a measure theoretical random variable $X$ with all finite moments, we can construct the algebraic probability space generated by $X$ (Example 3). Then through GNS representations, $X$ can be decomposed into algebraic random variables which are noncommutative each other. In particular, by using the notion known as "interacting Fock space" introduced in the Sect. 3.8, $X$ is identified (up to stochastic equivalence Definition 11) with the sum of "creation", "annihilation" and "preservation" operators, which satisfy a generalized form commutation relations (see [10], for details.). Such decomposition is called "quantum decomposition".

Then we can reconstruct various kinds of random variables from elemental noncommutative random variables. For example, in the *quantum harmonic oscillator*(or *1-mode Boson Fock space*), which will be introduced in the Sect. 3.7, a Gaussian random variable $X$ is represented as $X = a + a^*$ which satisfies $[a, a^*] = 1$. Interestingly, a Poisson random variable (with parameter $\lambda$) can be represented as $(A + \sqrt{\lambda})^*(A + \sqrt{\lambda})$ (see [10]).

## 3.7 Quantum-Classical Correspondence for Harmonic Oscillators

### *3.7.1 What is Quantum-Classical Correnspondence?*

*Quantum-classical correspondence* was the great guiding principle in the dawn of quantum theory to discover the hidden nature of phenomena in atomic scale. The quantum-classical correspondence principle as the guiding principle is, in short, can be stated as follows:

Principle: The quantum theory should recover the classical theory in the limit.

Ojima [8] has advocated the conceptual importance of the principle as the foundation of quantum theory and formulate it in the general scheme called *Micro-Macro duality* from the category the the universal framework (see [8], for example). In the present paper we focus on the most elementary example: Quantum-classical correspondence for *harmonic oscillator* and similar systems, where the probability measure called the *Arcsine law* plays a crucial role (Fig. 3.5).

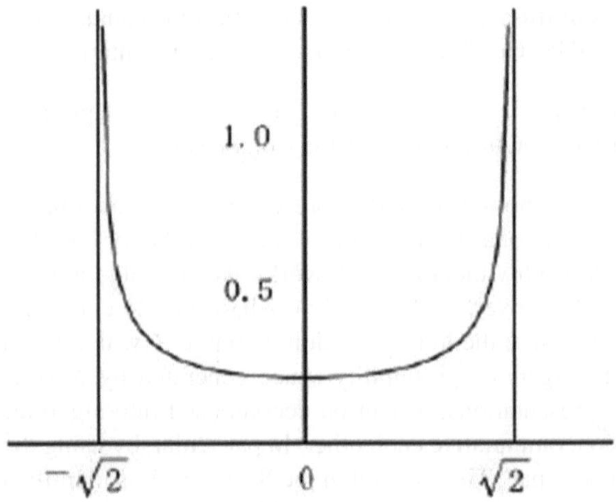


**Fig. 3.5** The Arcsine law

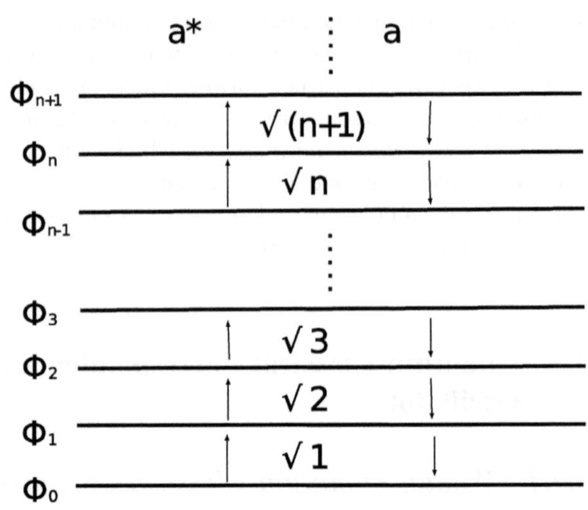


**Fig. 3.6** Quantum harmonic oscillator

### 3.7.2 *The Arcsine Law*

The (normalized) *Arcsine law* $\mu_{As}$ is the probability distribution on $\mathbb{R}$ with support $[-\sqrt{2}, \sqrt{2}]$ defined as

$$\mu_{As}(dx) = \frac{1}{\pi} \frac{dx}{\sqrt{2 - x^2}},$$

whose $n$th moment $M_n := \int_{\mathbb{R}} x^n \mu_{As}(dx)$ is given by

$$M_{2m+1} = 0, \quad M_{2m} = \frac{1}{2^m} \binom{2m}{m}.$$

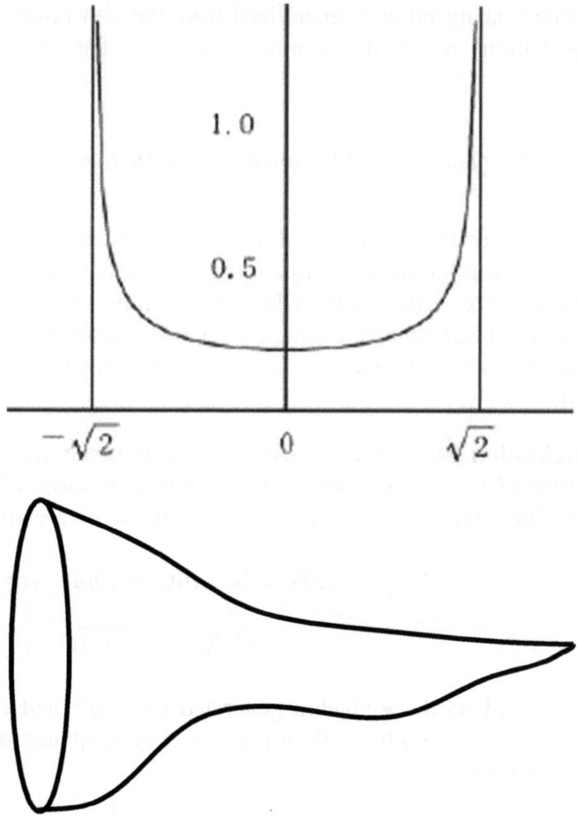


**Fig. 3.7** 1-dimensional density of DP: Fiber probe

In this case, the moment sequence $\{M_n\}$ characterizes $\mu_{As}$ ("the moment problem is deterministic").

The distribution $\mu_{As}$ often appears in probability theory. In the present paper we discuss another aspect of this distribution: the relationship with the notion of *harmonic oscillator.*

Let $x(t) = A \sin t$ be a harmonic oscillator with amplitude $A$, well known in classical mechanics. Then it is easy to show that "the time-averaged distribution $\mu$ of position $x$ " has the form

$$\mu(dx) = C \frac{dx}{\sqrt{A^2 - x^2}}$$

where $C$ denotes the normalizing constant, through elementary calculus. In $A = \sqrt{2}$ case (i.e. variance 1 case), $\mu = \mu_{As}$.

Here a natural question arises: Is it possible to see whether and in what meaning the *quantum-classical correspondence* holds for harmonic oscillators? This question, which is related to fundamental problems in Quantum theory and asymptotic analysis, is analyzed and generalized from the viewpoint of noncommutative algebraic probability with quite a simple combinatorial argument.

### 3.7.3 Quantum Harmonic Oscillator

First we review the notion of *quantum harmonic oscillator* (or 1-mode Boson Fock space) well known in the quantum mechanics, just in terms of quantum probability theory (Fig. 3.6). It is the GNS representation of so-called CCR (Canonical Commutation Relations) algebra, a $^*$-algebra generated by the elements $a, a^*$ satisfying the definitive relation $aa^* - a^*a = 1$. Or equivalently, we can define the notion as follows:

**Definition 19** (*Quantum harmonic oscillator*) A quantum harmonic oscillator is a triple $(\Gamma(\mathbb{C}), a, a^*)$ where $\Gamma(\mathbb{C})$ is a Hilbert space $\Gamma(\mathbb{C}) := \oplus_{n=0}^{\infty}\mathbb{C}\Phi_n$ with inner product given by $< \Phi_n, \Phi_m >= \delta_{n,m}$, and $a, a^*$ are operators defined as follows:

$$a\Phi_0 = 0, \quad a\Phi_n = \sqrt{n}\Phi_{n-1}(n \geq 1)$$

$$a^*\Phi_n = \sqrt{n+1}\Phi_{n+1}$$

Let $\mathcal{A}$ be the $*$-algebra generated by $a, a^*$, and $\varphi_n(\cdot)$ be the state defined as $\varphi(\cdot) := <\Phi_n, (\cdot)\Phi_n>$. Then $(\mathcal{A}, \varphi_n(\cdot))$ is an quantum probability space. It is well known that

$$X := a + a^*$$

represents the *position* observable and that

$$X \sim_{\varphi_0} \frac{1}{\sqrt{2\pi}} e^{-\frac{1}{2}x^2} dx.$$

That is, in $n = 0$ case, the distribution of position is Gaussian.

### 3.7.4 Quantum-Classical Correspondence for Harmonic Oscillator

On the other hand, the asymptotic behavior of the distributions of position as $n$ tends to infinity is nontrivial. In other words, what is the *classical limit* of the quantum harmonic oscillator? This question, which is related to fundamental problems in quantum theory and asymptotic analysis, was analyzed in [3] from the viewpoint of quantum probability with quite a simple combinatorial argument. The answer for this question is that the classical limit, which is defined later, for quantum harmonic oscillator is nothing but the Arcsine law.

**Theorem 5** ([3]) *Let $\Gamma_{\{\omega_n=n\},\{\alpha_n\equiv 0\}} := (\Gamma(\mathbb{C}), a, a^*)$ be the quantum harmonic oscillator, $X := a + a^*$ and $\mu_N$ be a probability distribution on $\mathbb{R}$ such that*

$$\frac{X}{\sqrt{2N-1}} \sim_{\varphi_N} \mu_N.$$

*Then $\mu_N$ weakly converges to the Arcsine law $\mu_{As}$.*

We will prove this in quite a general framework, called (1-mode) *interacting Fock spaces* which is defined in the next section.

## 3.8 Quantum-Classical Correspondence for Interacting Fock space

### *3.8.1 Interacting Fock Space*

**Definition 20** (*Jacobi sequence*) A sequence $\{\omega_n\}$ is called a Jacobi sequence if it satisfies one of the conditions below:

- (finite type) There exist a number $m$ such that $\omega_n > 0$ for $n < m$ and $\omega_n = 0$ for $n \geq m$;
- (infinite type) $\omega_n > 0$ for all $n$.

**Definition 21** (*Interacting Fock space*) Let $\{\omega_n\}$ be a Jacobi sequence and $\{\alpha_n\}$ be a real sequence. An interacting Fock space $\Gamma_{\{\omega_n\},\{\alpha_n\}}$ is a quadruple $(\Gamma(\mathbb{C}), a, a^*, a^o)$ where $\Gamma(\mathbb{C})$ is a Hilbert space $\Gamma(\mathbb{C}) := \oplus_{n=0}^{\infty}\mathbb{C}\Phi_n$ with inner product given by $\langle \Phi_n, \Phi_m\rangle = \delta_{n,m}$, and $a, a^*$ are operators defined as follows:

$$a\Phi_0 = 0, \quad a\Phi_n = \sqrt{\omega_n}\Phi_{n-1} (n \geq 1),$$

$$a^*\Phi_n = \sqrt{\omega_{n+1}}\Phi_{n+1},$$

$$a^o\Phi_n = \alpha_{n+1}\Phi_n.$$

Let $\mathcal{A}$ be the $*$-algebra generated by $\{a, a^*, a^o = (a^o)^*\}$, and $\varphi_n$ be the state defined as $\varphi_n(\cdot) := \langle \Phi_n, (\cdot)\Phi_n\rangle$. Then $(\mathcal{A}, \varphi_n)$ is a quantum probability space.

The notion of interacting Fock spaces is nothing but the generalization of quantum harmonic oscilator. It is the mathematical structure emerged out of the simplest model of (not-so strongly) interacting quantum fields [10, 11]. Moreover, it connects quantum probability theory to orthogonal polynomials of probability measures, as we will seen in the next subsection.

### 3.8.2 Interacting Fock Spaces and Orthogonal Polynomials for Probability Measures

Theorems for interacting Fock spaces often have interesting interpretation in terms of *orthogonal polynomials*. To see this we review the relation between interacting Fock spaces, probability measures and orthogonal polynomials. Let $\mu$ be a probability measure on $\mathbb{R}$ with all finite moments. (For the rest of the present paper, we always assume that all the moments are finite.) Then the space of polynomial functions is contained in the Hilbert space $L^2(\mathbb{R}, \mu)$. A Gram-Schmidt procedure provides orthogonal polynomials which only depend on the moment sequence.

Let $\{p_n(x)\}_{n=0,1,...}$ be the monic orthogonal polynomials of $\mu$ such that the degree of $p_n$ equals to $n$. Then there exist sequences $\{\alpha_n\}_{n=0,1,...}$ and Jacobi sequence $\{\omega_n\}_{n=1,2,...}$ such that

$$xp_n(x) = p_{n+1}(x) + \alpha_{n+1} p_n(x) + \omega_n p_{n-1}(x) \quad (p_{-1}(x) \equiv 0).$$

$\alpha_n \equiv 0$ if $\mu$ is symmetric, i.e., $\mu(-dx) = \mu(dx)$. We call $\{\{\omega_n\}, \{\alpha_n\}\}$ Jacobi sequences corresponding to $\mu$.

It is known that there exist an isometry $U : \Gamma_{\{\omega_n\}} \to L^2(\mathbb{R}, \mu)$ which sends $X := a + a^* + a^o$ to the multiplication operator, and through which we obtain

$$X := a + a^* + a^o \sim_{\varphi_N} |P_N(x)|^2 \mu(dx)$$

where $P_n$ denotes the normalized orthogonal polynomial of degree $n$ [10, 12].

In other words, through $U$, we can "decompose" a (measure-theoretic/classical) random variable into the sum of non-commutative algebraic random variables. (This crucial idea in quantum probability theory is called *quantum decomposition* [10].)

### 3.8.3 The Notion of a Classical Limit Distribution for an Interacting Fock Spaces

As the case for quantum harmonic oscillators, we define the notion of a *classical limit distributions* for an interacting Fock spaces. It is a distribution to which the distribution for $X$ under $\varphi_N$, after normalization, converges in moment.

**Definition 22** (*Classical Limit distribution*) Let $\Gamma_{\{\omega_n\},\{\alpha_n\}} := (\Gamma(\mathbb{C}), a, a^*, a^o)$ be an interacting Fock space, $X := a + a^* + a^o$ and $\mu_N$ be a probability distribution on $\mathbb{R}$ such that

$$\frac{X - \alpha_{N+1}}{\sqrt{\omega_N + \omega_{N-1}}} \sim_{\varphi_N} \mu_N.$$

A probability distribution $\mu$ on $\mathbb{R}$ is called a classical limit distribution if $\mu_N$ converge $\mu$ in moment.

By normalization, $\mu_N$ has mean 0 and variance 1.

*Remark 4* The uniqueness of classical limit distribution depends on the moment problem. Note that convergence in moment implies weak convergence in case the limit distribution is the solution of a determinate moment problem (see [13], for example.).

### 3.8.4 The Arcsine Law as Classical Limit Distribution

In the present subsection we prove that the Arcsine law is the classical limit distribution for interacting Fock spaces satisfying a general condition.

**Theorem 6** *Let $\Gamma_{\{\omega_n\},\{\alpha_n\}} := (\Gamma(\mathbb{C}), a, a^*, a^o)$ be an interacting Fock space satisfying*

$$\lim_{n\to\infty} \frac{\omega_{n+1}}{\omega_n} = 1, \quad \lim_{n\to\infty} \frac{\alpha_n - \alpha_{n+1}}{\sqrt{\omega_n + \omega_{n+1}}} = 0.$$

*Then the classical limit distribution is the Arcsine law $\mu_{As}$.*

*Proof* We will show that $\varphi_N((\frac{X-\alpha_{N+1}}{\sqrt{\omega_N+\omega_{N-1}}})^n) = \varphi_N((\frac{a+a^*+a^o-\alpha_{N+1}}{\sqrt{\omega_N+\omega_{N-1}}})^n)$ converge to $m$th moment of the Arcsine law. It is easy to show that

$$\frac{\omega_{N+k}}{\omega_N} \to 1 \quad (N \to \infty),$$

and the effect of $\frac{a^o-\alpha_{N+1}}{\sqrt{\omega_N+\omega_{N-1}}}$ tends to 0. Hence it suffices to calculate $\varphi_N((\frac{a+a^*}{\sqrt{\omega_N+\omega_{N-1}}})^n)$.

First, it is clear that

$$\varphi_N\left(\left(\frac{a+a^*}{\sqrt{\omega_N+\omega_{N-1}}}\right)^{2m+1}\right) = \langle \Phi_N, \left(\frac{a+a^*}{\sqrt{\omega_N+\omega_{N-1}}}\right)^{2m+1} \Phi_N\rangle = 0$$

since $\langle \Phi_N, \Phi_M\rangle = 0$ when $N \neq M$. To consider the moments of even degrees, we introduce the following notations:

- $\Lambda^{2m} := \{\text{maps from } 1, 2, \ldots, 2m \text{ to } \{1, *\}\}$,
- $\Lambda^{2m}_m := \{\lambda \in \Lambda^{2m}; |\lambda^{-1}(1)| = |\lambda^{-1}(*)| = m\}$.

Note that the cardinality $|\Lambda^{2m}_m|$ equals to $\binom{2m}{m}$ because the choice of $\lambda$ is equivalent to the choice of $m$ elements which consist the subset $\lambda^{-1}(1)$ from $2m$ elements in $\{1, 2, \ldots, 2m\}$.

It is clear that for any $\lambda \notin \Lambda^{2m}_m$

$$\langle \Phi_N, a^{\lambda_1} a^{\lambda_2} \cdots a^{\lambda_{2m}} \Phi_N\rangle = 0$$

since $\langle \Phi_N, \Phi_M\rangle = 0$ when $N \neq M$. On the other hand, for any $\lambda \in \Lambda^{2m}_m$

$$\frac{1}{\omega_N^m}\langle \Phi_N, a^{\lambda_1}a^{\lambda_2}\cdots a^{\lambda_{2m}}\Phi_N\rangle \to 1 \quad (N\to\infty)$$

holds since $\langle \Phi_N, a^{\lambda_1}a^{\lambda_2}\cdots a^{\lambda_{2m}}\Phi_N\rangle$ becomes the product of $2m$ terms having the form $\sqrt{\omega_{N+k}}$ ($k$ is an integer and $-m+1\leq k\leq m$) and

$$\frac{\omega_{N+k}}{\omega_N} \to 1 \quad (N\to\infty)$$

as we have mentioned. Hence,

$$\varphi_N\left(\left(\frac{a+a^*}{\sqrt{\omega_N+\omega_{N-1}}}\right)^{2m}\right) = \left\langle \Phi_N, \left(\frac{a+a^*}{\sqrt{(1+\frac{\omega_{N-1}}{\omega_N})\omega_N}}\right)^{2m}\Phi_N\right\rangle$$

$$= \frac{1}{\left(1+\frac{\omega_{N-1}}{\omega_N}\right)^m}\sum\nolimits_{\lambda\in\Lambda^{2m}}\frac{1}{\omega_N^m}\langle \Phi_N, a^{\lambda_1}a^{\lambda_2}\cdots a^{\lambda_{2m}}\Phi_N\rangle$$

$$= \frac{1}{\left(1+\frac{\omega_{N-1}}{\omega_N}\right)^m}\sum\nolimits_{\lambda\in\Lambda^{2m}_m}\frac{1}{\omega_N^m}\langle \Phi_N, a^{\lambda_1}a^{\lambda_2}\cdots a^{\lambda_{2m}}\Phi_N\rangle$$

$$\to \frac{1}{2^m}|\Lambda^{2m}_m| = \frac{1}{2^m}\binom{2m}{m} \quad (N\to\infty).$$

*Remark 5* Since the Arcsine law is the solution of a determinate moment problem, moment convergence implies weak convergence. (see [13], for example.)

The theorem means that the Arcsine law $\mu_{As}$ is turned out to be the classical limit distribution of many interacting Fock spaces. For example, interacting Fock spaces corresponding to uniform distribution, exponential distribution or "q-Gaussians" ($-1<q\leq 1$, $\omega_n=[n]_q := 1+q+q^2+\cdots+q^{n-1}$, $\alpha_n\equiv 0$. $q=1$ is Gaussian and $q=0$ is Wigner Semicircle Law) satisfy the condition above. For more general condition, we have the "discretized version" of the Arcsine law (see [4]).

### 3.8.5 An Asymptotic Behavior of Orthogonal Polynomials

The theorem above implies description of an asymptotic behavior of orthogonal polynomials:

**Corollary 1** *Let $\mu$ be a prabablity measure such that the corresponding Jacobi sequences $\{\omega_n\}$, $\{\alpha_n\}$ satisfies the conditions above,*

*Then the measure $\mu_n$ defined as $\mu_n(dx) := |P_n(\sqrt{\omega_N+\omega_{N-1}}x)|^2\mu(\sqrt{\omega_N+\omega_{N-1}}dx)$ weakly converge to $\mu_{As}$.*

Many kinds of orthogonal polynomials such as Jacobi polynomials(e.g. Legendre polynomials), Laguerre polynomials or $q$-Hermite polynomials($-1 < q \leq 1$) satisfy the condition above. For more general condition, see [4].

## 3.9 Toward Applications for Dressed Photons

### *3.9.1 Arcsine Law as the Distribution of DP*

As a result of preceding sections, the following fact is established:

> The quantum harmonic oscillator and its analogues (i.e. interacting Fock spaces) on the states corresponding to higher quantum number behave very much like as the classical harmonic oscillator.

As is well known, the light-matter composite system can be considered as (analogues of) quantum harmonic oscillator. By the our definition of DP, the below follows directly:

> The time averaged distribution for the position of DP are considered as the (normalized) Arcsine law, when normalized.

Note that by the interaction between the light and matter, each mode of DP has effective mass and it is not paradoxical to consider the position of DP. Moreover, when the spatical extension of matter can be compactly supported, the variance will be determined by the length of the matter. Then we have essentially one distribution: The normalized Arcsine law.

Let us remark about the role of photon here. The strength of interaction with photon results in the higher/lower distinction. If the stronger the interaction, the higher the suitable boundary between "higher family"=DP and "lower family"=heat bath. Hence, in general, when the interaction is enough strong the distribution of DP will be close to the Arcsine law.

As the shape of Arcsine law are polalized, extreme point of the matter will have the highest (1-dimensional) density. That is,

> The probability distribution of DP will be the densest around the extreme points of the matter.

That is why the localization of DP occurs around boundary universally. This is quite coherent with the many experiments [1]. We had given the simple explanation for the this universality by the method of quantum probability theory.

### *3.9.2 Designing Fiber Probe*

One of the most fundamental example of the localization of DP around the end point of the fiber probe [1]. As the collorary of the discussions in the previous secion, the

1-dimensional density of DP is high in the both extreme point (Fig. 3.7): The point of limitation by wavelength and the tip of the fiber probe.

Then the 3-dimensional density, when normalize to the variance 1, will be the following:

$$\mu_{vol}(dx) = C\frac{dx}{S(x)\sqrt{2-x^2}}$$

Where $S(x)$ is the cross sectional Area and the $C$ is the normalizing constant to make the measure to be probability measure.

As a result, the more the rapid the decreasing of $S(x)$, the more dense the concentration of DP. This will become the principle of the *design* of the fiber probe. On the other hand, the dissipativity is still to be considered.The author and colleagues are trying to calculate "the best" shape $S(x)$ of the fiber probe based on the variational method combining these viewpoints.

### *3.9.3 Impurity*

The last thing to discuss in the present paper is the role of impurity in matter. For instance, let us consider the case that the fiber probe has impurity. Such impurity play the role of nano-scale "cavity" for the DP, and then in that nano-cavity there should be again the micro Arcsine-like distribution.

Another example is the photon breeding by DP [14]. It is also explained by the impurity effect, and coherent with our viewpoint based on the Arcsine law as far. Moreover, taking into the polarized proberty of the Arcsine law again, our model will leads to the new prediction:

> If our model is correct, then the same kind of impurity will cause different amount of the light emission, depending on the place of the impurity: The impurity near boundary may cause stonger emission than the similar one far away from boundary.

The author would know whether this prediction is testable by experiments, and then whether it is true or not. We will discover, the new phenomena or leads to the deeper reflection on the hidden important hypothesis on the our picture, anyway. For the quantum probability approach is quite abstract and general, the prediction become universal. This universality is nothing but we seek in the present paper.

## 3.10 Conclusion

The quantum probability methods predict the following universal phenomena:

> The high concentration of DP occur at the extreme points and around the points of impurity.

In short:

Localization of DP occurs around singularity .

The slogan not only lead to the fundamental issues on DP, but it will open the way to the interplay between the study of DP and various fertile branches of mathematics. The author strongly expect that DP will be connected to all over the sciences through QP.

**Acknowledgements** The author is grateful to Dr. Motoichi Ohtsu, Dr. Izumi Ojima and Dr. Hiroki Sako for the fruitful collaboration. He also thanks to Dr. Hirofumi Sakuma, Dr. Itsuki Banno, Dr. Suguru Sangu, Dr. Hiroshi Ando and Dr. Kazuya Okamura for important discussions. This work was supported by Research Origin for Dressed Photon.

## Appendix A: $C^*$-Probability Space

### $C^*$-Probability Space

**Definition 23** An algebra $\mathcal{A}$ with a norm $\|\cdot\|$ is said to be a normed algebra if

$$\|ab\| \leq \|a\|\|b\|,$$

holds for all $a, b \in \mathcal{A}$. When $\mathcal{A}$ is a Banach space (i.e.,complete with respect to $\|\cdot\|$), it is said to be a Banach algebra.

**Definition 24** A $^*$-algebra $\mathcal{A}$ with norm $\|\cdot\|$ is said to be a $C^*$-algebra if the following hold:

(i) $\mathcal{A}$ is a Banach algebra with respect to $\|\cdot\|$
(ii) $\|a^*a\| = \|a\|^2$.

**Proposition 3** *If $\mathcal{A}$ is unital, then we obtain $\|a^*\| = \|a\|$ and $\|1\| = 1$.*

**Definition 25** A quantum probability space $(\mathcal{A}, \varphi)$ is said to be a $C^*$-probability space when $\mathcal{A}$ is unital $C^*$-algebra. If $\mathcal{A}$ is commutative, $(\mathcal{A}, \varphi)$ is said to be commutative $C^*$-probability space.

A measure theoretic probability spaces can be viewed as classical probability spaces. Moreover, a classical probability space is isomorphic to the associated conventional (measure-theoretical) probability space when it is a commutative $C^*$-algebraic space. This is a consequence of the representation theorem for commutative $C^*$-algebra due to Gel'fand and Naimark, discussed below. The content of the theorem is simple and impressive: The notion of unital commutative $C^*$- algebra is the "dual" of the notion of compact Hausdorff space.

### Commutative $C^*$-Probability Spaces

**Definition 26** Let $\mathcal{A}$ be a commutative Banach algebra. A nonzero linear functional $\omega : A \longrightarrow \mathbb{C}$ is said to be a character on $\mathcal{A}$ if it satisfies

$$\omega(ab) = \omega(a)\omega(b)$$

for all $a, \in \mathcal{A}$. (That is, $\omega$ is homomorphism of ring.) The set of all characters on $A$ is called the character space of $\mathcal{A}$ and denoted as $\Gamma_{\mathcal{A}}$.

When $\mathcal{A}$ is unital, $\Gamma_{\mathcal{A}}$, which is naturally identified with the set of all maximal ideals in $\mathcal{A}$, becomes a compact space (with respect to weak* topology). Moreover, any functional $\hat{a}$ defined by

$$\hat{a}(\omega) := \omega(a)$$

becomes a continuous function on $\Gamma_{\mathcal{A}}$.

**Definition 27** The homomorphism $\hat{}: \mathcal{A} \longrightarrow C(\Gamma_{\mathcal{A}})$ is called Gel'fand transform of unital commutative Banach algebra $\mathcal{A}$.

Summing up, we obtain the following theorem:

**Theorem 7** (Gel'fand-Naimark) *A unital commutative Banach algebra $\mathcal{A}$ is represented by $C(\Gamma_{\mathcal{A}})$ thorough the Gel'fand transform.*

**Corollary 2** *For unital commutative Banach algebra $\mathcal{A}$,*

$$\hat{a}(\Gamma_{\mathcal{A}}) = Spec_{\mathcal{A}}(a) := \{\lambda \in \mathbb{C}; \lambda 1 - a \ \textit{ is not invertible in } \mathcal{A}\}$$

*holds for all $a \in \mathcal{A}$.*

By the fact above, $\Gamma_{\mathcal{A}}$ is also called the "spectrum of $\mathcal{A}$" and often denoted as $Spec_{\mathcal{A}}$, for example.

As $C(\Gamma_{\mathcal{A}})$ becomes a unital commutative $C^*$-algebra equipped with the involution $f^*(\omega) := \overline{f(\omega)}$ and the norm $\|f\| := \sup_{\omega \in \Gamma_{\mathcal{A}}} \{|f(\omega)|\}$, Gel'fand–Naimark theorem implies that any unital commutative Banach algebra $\mathcal{A}$ can be represented by unital commutative $C^*$-algebra. If $\mathcal{A}$ itself is unital commutative $C^*$-algebra, the following theorem holds:

**Theorem 8** (Gel'fand-Naimark) *A unital commutative $C^*$-algebra $\mathcal{A}$ is isometrically isomorphic to $C(\Gamma_{\mathcal{A}})$ as $C^*$-algebras through the Gel'fand transform. Moreover, $\mathcal{A} \cong C(\Omega)$ implies $\Gamma_{\mathcal{A}} \cong \Omega$ when $\Omega$ is compact.*

As a result, the category of unital commutative $C^*$-algebras and the category of compact Hausdorff spaces are contravariantly equivalent. That is, the notion of commutative $C^*$-algebra is the "dual" of the notion of compact Hausdorff space.

*Remark 6* If we just replace "compact" by "locally compact", "$C(X)$" by "$C_0(X)$ (continuous functions which vanish at infinity)" and "unital $C^*$-algebra" by "$C^*$-algebra, the main results above hold. Then we can see that Gel'fand transforms are nothing but "generalized Fourier transform": Moreover, the results above also provide the basis for "functional calculus" for operators.

For the application of Gel'fand–Naimark theorem for noncommutative probability, we use the theorem below:

**Theorem 9** (Riesz–Markov–Kakutani) *Let $\Omega$ be a compact Hausdorff space. There is one-to-one correspondence between $\{\varphi;\ \varphi$ is a positive linear functional on $C(\Omega)\}$ and $\{\mu;\ \mu$ is a finite regular Borel measure$\}$ through the integration below:*

$$\varphi(f) = \int_{\Omega} f(\omega)\mu(d\omega)$$

*In particular, there is one-to-one affine correspondence between states on $C(\Omega)$ and regular Borel probability measures. (A Borel measure $\mu$ is said to be regular if*

$$\mu(B) = \sup\{\mu(K);\ K \subset B,\ K \text{ is compact}\} = \inf\{\mu(O) : B \subset O,\ O \text{ is open}\}$$

*holds for any Borel set $B$ in $\Omega$.)*

**Theorem 10** *Any commutative $C^*$-probability space is represented by a measure-theoretical probability space $(\Omega, \mathfrak{B}(\Omega), P)$, where $\Omega$ is a compact Hausdorff space and $P$ is a regular Borel probability measure on the Borel algebra $\mathfrak{B}(\Omega)$.*

The representation Theorems 8 and 9 have a lot of applications. One of the typical application to general (not necessarily commutative) $C^*$-probability space is in the proof for the theorem below on the boundedness (continuity) of the state.

**Theorem 11** *Let $(\mathcal{A}, \phi)$ be a $C^*$-probability space. Then the state $\varphi$ is a bounded linear functional such that*

$$|\varphi(a)| \leq \|a\|$$

*for all $a \in \mathcal{A}$, and hence, $\|\varphi\| = 1$.*

*Proof* By Theorem 2,

$$|\varphi(a)|^2 = |\varphi(1 \cdot a)|^2 \leq \varphi(1)\varphi(a^*a) = \varphi(a^*a)$$

holds for all $a \in \mathcal{A}$. On the other hand, applying the Gel'fand representation theorem for the unital commutative algebra generated by $a^*a$, we obtain $a^*a \leq \|a\|$ as an obvious inequality for functions. Hence

$$\varphi(a^*a) \leq \varphi(\|a\|^2) = \|a\|^2\varphi(1) = \|a\|^2$$

holds for all $a \in \mathcal{A}$ by Theorems 8, 9 and Proposition 3. Combining the above two inequalities, we obtain the result.

## Appendix B: Moments in Probability Theory

### *Moments of Famous Distributions*

**Definition 28** Let $X$ be a random variable and $E(X)$ denotes the expectation (mean) of $X$ (if it exists). Then $E(X^n)$ is called the $n$th moment of $X$ (if it exists) and denoted as $M_n(X)$.

*Remark 7* There exist many kinds of important random variables which have no finite moments. For example, a Cauchy random variable has no mean. From now on, however, we will concentrate on random variables which admit all finite moments.

*Example 4* Let us consider the Poisson random variable $X$ with parameter $\lambda$, that is, its probability distribution $\mu_\lambda$ is given by the Poisson law

$$\mu_\lambda = e^{-\lambda} \sum_{n=0}^{\infty} \frac{\lambda^n}{n!} \delta_n .$$

For the $\lambda = 1$ case (normalized to variance 1), the $n$th moments $M_n(X)$ is equal to $n$th Bell number:
$\mathcal{B}_n :=$ (the number of all quotient sets of a set with $n$ elements).

Note that $E((X)_n) = 1$ is rather easy to prove, where $(x)_n$ denotes the falling factorial $x(x-1)(x-2)\cdots(x-n+1)$.

Let us discuss the moments of two distributions, which play quite important roles in noncommutative probability, as examples.

*Example 5* Let $X$ be a Gaussian random variable whose probability measure $\mu$ is given by normalized Gaussian

$$\mu(dx) = \frac{1}{\sqrt{2\pi}} e^{-\frac{x^2}{2}} dx .$$

Then

$$M_{2m}(X) = \frac{(2m)!}{2^m m!} = (2m-1)!! , \quad M_{2m-1}(X) = 0$$

*Example 6* Let $X$ be a random variable whose probability density function is given by

$$\rho(x) = \begin{cases} \frac{1}{2\pi}\sqrt{4-x^2} \ , |x| \leq 2 \\ 0 \ , |x| > 2 \end{cases}$$

The probability measure defined above is called normalized *Wigner semicircle law*. Then its moments are

$$M_{2m} = C_m = \frac{(2m)!}{(m+1)!m!} , \quad M_{2m-1} = 0.$$

Here $C_m$ is known as $m$th Catalan number, which has various interesting combinatorial meaning. (For example, see [15]).

Of course the moments above should have some combinatorial meaning: The present author found a combinatorial interpretation in [3] which is quite essential to give a simple proofs for the theorem related to quantum probability.

## *Moment Problem*

In a previous sections we have considered the moments as the characteristics of probability measure. When the moments characterizes distribution uniquely? A simple and celebrated sufficient condition by Carleman is stated as follows:

**Theorem 12** *Suppose that the moment sequence* $M_0, M_1, M_2, \ldots$ *satisfies the condition*

$$\sum_{m=1}^{\infty} M_{2m}^{-\frac{1}{2m}} = \infty$$

*Then the moment sequence characterize probability measure uniquely,i.e., there is only one probability measure whose moment sequence is* $M_0, M_1, M_2, \ldots$.

When the moment sequence characterizes probability uniquely, it is said that the moment problem is determinate, and such unique probability measure is said to be a solution for the determinate moment problem.

**Corollary 3** *If* $M_n \le n^n$, *the moment problem is determinate.*

*Remark 8* As $n^n$ is nothing but the number of mappings from a set with $n$-elements to itself (or another set with $n$-elements), cardinals of the sets consisting of partitions satisfy the condition above.

Then, Poisson probability measure and Gauss/semicircle/Bernoulli probability measure is uniquely characterized by its moments.

*Remark 9* When a probability measure has compact support in $[-L, L]$, it is easy to see that it is a solution for the determinate moment problem because

$$M_{2m} \le \int_{[-L,L]} x^{2m} \mu(dx) \le L^{2m}.$$

From the discussion above, the moment problem for the Arcsine law is also determinate.

Moreover, convergence of moments characterizes weak convergence if the moment problem for limit moments are deterministic.

**Theorem 13** *Convergence of moments implies weak convergence of probability measures, if the limit measure is a solution for the moment problem; one need not assume that the probability measures in the sequences are solutions for the determinate moment problem. (see Theorem 4.5.5 in [13]).*

## References

1. M. Ohtsu, *Dressed Photons* (Springer, Berlin, 2014)
2. H. Saigo, From Combinatorics to Noncommutative Probability. DR thesis, Kyoto University, Kyoto, 2011
3. H. Saigo, Infin. Dimens. Anal. Quantum Probab. Relat. Top. **15**(3), 1250021 (2010)
4. H. Saigo, H. Sako, Ann. Inst. Henri Poincaré Comb. Phys. Interact. **3**, 405 (2016)
5. T.D. Newton, E.P. Wigner, Rev. Mod. Phys. **21**, 400 (1949)
6. A. Wightman, Rev. Mod. Phys. **34**, 845 (1962)
7. H. Saigo, M. Ohtsu, I. Ojima, Appl. Phys. A **123**, 724 (2017)
8. I. Ojima, in *Proceedings of the International Conference of Stochastic Analysis*, ed. by T. Hida (World Scientific, Singapore, 2005)
9. A. Kolmogorov, *Grundbegriffe der Wahrscheinlichkeitsrechnung* (Springer, Berlin, 1933)
10. A. Hora, N. Obata, *Quantum Probability and Spectral Analysis of Graphs* (Springer, Berlin, 2007)
11. L. Accardi, N. Obata, *Foundations of Quantum Probability* (in Japanese) (Makino Shoten Publisher, Tokyo, 2003)
12. L. Accardi, M. Bożejko, Infin. Dimens. Anal. Quantum Probab. Relat. Top. **1**, 663 (1998)
13. K.L. Chung, *A Course in Probability Theory* (Academic, San Diego, 1968)
14. M. Ohtsu, *Silicon Light Emitting Diodes and Lasers* (Springer, Berlin, 2016)
15. J.H. Conway, R.K. Guy, *The Book of Numbers* (Copernicus, New York, 1996)

# Chapter 4
# Control over Off-Shell QFT via Induction and Imprimitivity

**Izumi Ojima**

**Abstract** In the framework of Micro–Macro duality with quadrality scheme, the mutual relations between Micro and Macro are considered in a systematic way. On this basis, many interesting aspects of symmetry breaking are discussed, according to which the classifying space of sectors is shown to have the structure of symmetric spaces characterized by the condition similar to Maxwell and Einstein equations. Applying these results to the attempts for constructing a theory of dressed photons, we are naturally led to the idea of the relevance of automorphic forms which have close connections with number theory.

## 4.1 Micro–Macro Duality + Quadrality Scheme

When we integrate a mathematical approach to the dynamical aspects of the system in question, together with the geometric description of the relevant structure in terms of invariants which are generated by the dynamical processes and which implement the classification of the processes and structures, then we arrive at the category-theoretical framework of "Micro–Macro duality + quadrality scheme" ([1]; "Quantum Fields and Micro–Macro Duality [2], a book by Ojima, 2013 in Japanese and also see [3]) by incorporating categorically the natural duality between the dynamical processes and the classifying spaces.

By analyzing closely in this framework the dynamical processes and the classifying scheme based on geometric invariants generated by the former processes, we can understand that both of the invisible Micro doman corresponding to the dynamical processes and of the visible Macro structure to the classifying structure in terms of geometric invariants constitute the duality structure, to be called "***Micro–Macro duality***" [4]. According to such general feature, the description in terms of momenta $p$ corresponds to the inivisible Micro domain and the familiar concept of "spacetime

I. Ojima (✉)
Research Origin for Dressed Photon, c/o Nichia Corp., 3-13-19 Moriya-cho, Kanagawa-ku, Yokohama, Kanagawa 221-0022, Japan
e-mail: ojima@za.ztv.ne.jp

 
T. Yatsui (ed.), *Progress in Nanophotonics 5*, Nano-Optics and Nanophotonics,
https://doi.org/10.1007/978-3-319-98267-0_4

points" is seen to be realized in the $x$-picture as visible Macro geometric structure formed in emergence processes from the microscopic motions.

The essence of the duality between on-shell $\leftrightarrows$ off-shell should be understood in this context as the on-shell corresponding to the particle-like Macro and the off-shell to the existence of quantum fields in virtual invisible modes before visualization into the forms of particles.

## 4.2 Quadrality Scheme $p_\mu$ Vs. $x^\nu$ and 5W1H

Microscopic processes of motions can be described by a group(oid) structure which drives the dynamical processes and which acts on the algebras of physical quantities, Macroscopic classifying structure emerging from the dynamical processes can be extracted from the structure of state space as the dual of the algebra of physical quantities and a geometric space emerges consisting of classifying indices extracted from states which functions as the dual of the Micro dynamical system. Putting altogether these four ingredients of dynamics, algebras, states and classifying space, they constitute a "***quadrality scheme***" describing "***Micro–Macro duality***" [4]:

| | | |
|---|---|---|
| $\nearrow$ | ***Classifying Space*** <br> = ***Spec*** | |
| (***Family of***) <br> ***States*** | $\leftrightarrows \upharpoonleft$ (***Representations***) $\downharpoonright$ $\leftrightarrows$ | ***Algebra*** |
| | ***Dynamics*** | $\nearrow$ |

The constituents appearing in this scheme can be understood as the standard components of "5W1H" required for the descriptions of events to be conveyed to other persons:

***Classifying Space***:
[When and where? as the vocabularies to specify spacetime localization of events],
(***Family of***) ***States***:
["Who?" = "Subject" to choose, indicate the context of description],
***Algebra*** (of variables for description):
[What?= specification of objects to be described],
(***Representations*** (of algebra)):
[How?= description of the modes of phenomena= "representation modules"],
***Dynamics***:
[Why?= causes, motives driving a process = dynamical laws].

## 4.3 Physical Nature Described as Dynamical Systems

The theoretical composition of the standard branches in modern physics can be understood as "physical nature from the viewpoint of dynamical systems", which implies that we describe the dynamical behaviours of physical nature on the basis of [dynamics of a physical system and the system of physical quantities driven by it: $G \underset{\alpha}{\curvearrowright} \mathcal{X}$].

$G$ is a group to describe the dynamics such as $G = \mathbb{R}$ or $\mathbb{Z}$, for example: i.e., time forming an additive group;

$\mathcal{X}$ denotes the algebra of physical quantities which is equipped with addition, muliplication and scalar multiplication:

$$\mathcal{X} \times \mathcal{X} \ni (A, B) \mapsto A + B, AB \in \mathcal{X},$$

$$K \times \mathcal{X} \ni (c, A) \mapsto cA \in \mathcal{X}.$$

As the scalar field $K$ we can suppose, in many cases, $K = \mathbb{R}$: real number coefficients. For the theoretical treatment of quantum theory, we need $K = \mathbb{C}$ complex number coefficients. For discussing such discrete structures as crystals the choise of finite fields may be convenient, but we start from $K = \mathbb{R}$ or $\mathbb{C}$ just for fixing the notation.

To give the precise meaning to such expression as "physical nature from the viewpoint of dynamical systems", we should choose a group action $\alpha : G \times \mathcal{X} \ni (s, A) \mapsto \alpha_s(A) \in \mathcal{X}$ as one of the most essential ingredients, which is supposed to satisfy

$$\alpha_s(c_1 A_1 + c_2 A_2) = c_1 \alpha_s(A_1) + c_2 \alpha_s(A_2),$$

$$\alpha_{st}(A) = \alpha_s(\alpha_t(A)),$$

where the group composition $G \times G \ni (s, t) \mapsto st \in G$ in $G$ is denoted generically as $st$ instead of $s + t$. When the combination of the structure of an algebra $\mathcal{X}$ and a group actrion $\alpha$ on $\mathcal{X}$ satisfies the above property, it is called a "dynamical system" and denoted in short by $G \underset{\alpha}{\curvearrowright} \mathcal{X}$.

## 4.4 Emergence of Sector Classifying Space *Spec*

In this mathematical framework for describing the process of emergence, the concept of a "sector" plays crucial roles, and hence, we need its precise definition.

What is a ***sector?*** for the mathematical description of a quantum system, we need a ***non-commutative*** **(*C**-)*algebra*** $\mathcal{X}$ (: ***Algebra***) of physical variables to characterize the system and a certain family of ***states*** $\omega \in E_{\mathcal{X}}$ to quantify the measured values

$\omega(A)$ of physical variables $A \in \mathcal{X}$. According to Gel'fand–Naimark–Segal theorem [5], a representation $(\pi_\omega, \mathfrak{H}_\omega, \Omega_\omega)$ (called GNS representation) of $\mathcal{X}$ is so constructed from $\omega$ that physical variables $A \in \mathcal{X}$ are represented as linear operators $\pi_\omega(A)$ acting on a Hilbert space $\mathfrak{H}_\omega$, the totality of which determines a very important concept of representation von Neumann algebra $\pi_\omega(\mathcal{X})'' =: \mathcal{X}_\omega$. Elements $C \in \mathfrak{Z}_\omega(\mathcal{X})$ of the center $\mathfrak{Z}_\omega(\mathcal{X})$ of $\mathcal{X}_\omega$ defined by

$$\mathfrak{Z}_\omega(\mathcal{X}) := \pi_\omega(\mathcal{X})'' \cap \pi_\omega(\mathcal{X})' = \mathcal{X}_\omega \cap \mathcal{X}_\omega',$$

are commuting with all elements $X$ in $\mathcal{X}_\omega$:

$$[C, X] = 0 \text{ for } \forall X \in \mathcal{X}_\omega$$

and play the role of "order parameters" as Macroscopic commutative observables. This commutativity allows the simultaneous diagonalization of $\mathfrak{Z}_\omega(\mathcal{X})$, which is, mathematically speaking, just the spectral decomposition of the commutative algebra $\mathfrak{Z}_\omega(\mathcal{X}) = L^\infty(Spec)$ with the spectrum of $\mathfrak{Z}_\omega(\mathcal{X})$ denoted by $Spec := Sp(\mathfrak{Z}_\omega(\mathcal{X}))$. The diagonalized situation with all the order parameters specified corresponds physically to a ***pure phase***, or mathematically corresponding to a quasi-equivalence class of a ***factor state*** $\gamma$ with a trivial center: $\mathfrak{Z}_\gamma(\mathcal{X}) = \mathcal{X}_\gamma'' \cap \mathcal{X}_\gamma' = \mathbb{C}1$, which is called a ***sector***. Here ***quasi-equivalence*** [6] means the ***unitary equivalence up to multiplicity*** and a factor state corresponds to a minimal unit of states or representations in the sense that its center cannot be decomposed any more.

***Warning!***: In usual quantum mechanics with ***finite*** degrees of freedom, we encounter such a special situation that there exists ***only one sector*** in any such theories, owing to the Stone–von Neumann uniqueness theorem [7] asserting the uniqueness of irreducible representations of the algebras of canonical commutation relations. In this situation, the Macro level consists of only one point, or more appropriately, ***the Macro level in such theories is absent***!! Since the relation between invisible Micro and visible Macro necessarily requires the commitment of the infinite degrees of freedoms, we cannot attain meaningful explanations about any of such "mysteries of quantum paradoxes" as "Schrődinger's cat", by sticking to the familiar quantum theory with ***finite*** degrees of freedom. This essential point has never properly been understood in the common discussions prevailing among "quantum physics industry" and science journalism.

## 4.5 Sectors and Disjointness

To understand properly the concept of sectors, it is crucial to note the following points about the mutual relations between different sectors. Namely, the relation between two ***different sectors*** $\pi_1, \pi_2$ is expressed by the concept of ***disjointness*** as follows:

$$T\pi_1(A) = \pi_2(A)T \ (\forall A \in \mathcal{X}) \implies T = 0,$$

which is stronger than unitary inequivalence and has deep implications as seen later. The macroscopic quantities characterized by their commutativity appear as the center $\mathfrak{Z}_\omega(\mathcal{X})$ of a mixed phase algebra $\pi_\omega(\mathcal{X})'' = \mathcal{X}_\omega$ containing many different sectors as pure phases, and its spectrum $Spec = Sp(\mathfrak{Z}_\omega(\mathcal{X}))$ as realized values $\chi \in Spec$ of order parameters $C \in \mathfrak{Z}_\omega(\mathcal{X})$ discriminates the pure phases contained in the mixed phase state $\omega$, The sectors as pure phases play the roles as the Micro–Macro boundary between microsocpic quantum system and macroscopic classical system as the environment, and they unify, at the same time, both these into a Micro–Macro composite system as a mixed phase. According to this story, the duality between intra-sectorial domains versus inter-sectorial relations holds as follows:

| ← ***Visible Macro*** consisting of ***sectors*** → | | | | ***inter-sectorial*** relations |
|---|---|---|---|---|
| $\cdots \quad \gamma_N$ | ***sectors*** $\gamma$ | $\gamma_2$ | $\gamma_1$ | $Spec$ |
| $\vdots$ | $\vdots$ | $\vdots$ | $\vdots$ | ↑ ***intra-sectorial*** |
| $\cdots \pi_{\gamma_N} \vdots$ | $\pi_\gamma \vdots$ | $\pi_{\gamma_2} \vdots$ | $\pi_{\gamma_1} \vdots$ | ‖ |
| $\vdots$ | $\vdots$ | $\vdots$ | $\vdots$ | ↓ ***invisible Micro*** |

The concept of sectors defined in this way as Micro–Macro boundaries between invisible Micro and visible Macro realizes the theoretical framework of quadrality scheme which provides the precise formulation of "quantum-classical correspondence".

## 4.6 Disjointness Vs. Quasi-equivalence

Along this line, we clarify the homotopical basis of Tomita theorem of central decomposition of states and representations [5].

In the C*-category $Rep_{\mathcal{X}}$ of representations of a C*-algebra $\mathcal{X}$, there exists the *universal representation* $\pi_u = (\pi_u, \mathfrak{H}_u) \in Rep_{\mathcal{X}}$ containing $\forall \pi = (\pi, \mathfrak{H}_\pi) \in Rep_{\mathcal{X}}$ as its subrepresentation: $\pi_u \succeq \pi = (\pi, \mathfrak{H}_\pi) \in Rep_{\mathcal{X}}$.

Such $\pi_u$ can be concretely realized as the direct sum $(\pi_u, \mathfrak{H}_u) := \bigoplus_{\omega \in E_{\mathcal{X}}} (\pi_\omega, \mathfrak{H}_\omega)$ of all the GNS representations, with the action of universal enveloping von Neumann algebra

$$\mathcal{X}'' \cong \mathcal{X}^{**} \cong \pi_u(\mathcal{X})'' \curvearrowright \mathfrak{H}_u.$$

For a representation $\pi \in Rep_{\mathcal{X}}$ its "disjoint complement" $\pi^{\overset{|}{\circ}}$ is defined as maximal representation disjoint from $\pi$:

$$\pi^{\overset{|}{\circ}} := \sup\{\rho \in Rep_{\mathcal{X}}; \rho \overset{|}{\circ} \pi\},$$

where the disjointness means $\rho \overset{|}{\circ} \pi \iff Rep_{\mathcal{X}}(\rho \to \pi) = \{0\}$: i.e., no non-zero intertwiners.

Then, we observe the following four points, (i)–(iv) [8]:

(i) $P(\pi^{\overset{|}{\circ}}) = c(\pi)^{\perp}$,

$$P(\pi^{\overset{|\,|}{\circ\circ}}) = c(\pi)^{\perp\perp} = c(\pi) := \bigvee_{u \in \mathcal{U}(\pi(\mathcal{X})')} u P_{\pi} u^{*} \in \mathcal{P}(\mathfrak{Z}(W^{*}(\mathcal{X}))),$$

where $P(\pi) \in W^{*}(\mathcal{X})'$ is defined as the projection corresponding to $(\pi, \mathfrak{H}_{\pi})$ in $\mathfrak{H}_{u}$ and $c(\pi)$ is the central support of $P(\pi)$ defined by the minimal central projection majorizing $P(\pi)$ in the center $\mathfrak{Z}(W^{*}(\mathcal{X})) := W^{*}(\mathcal{X}) \cap W^{*}(\mathcal{X})'$ of $W^{*}(\mathcal{X})$.

(ii) $\pi_1^{\overset{|\,|}{\circ\circ}} = \pi_2^{\overset{|\,|}{\circ\circ}} \iff \pi_1 \approx \pi_2$ (: quasi-equivalence= unitary equivalence up to multiplicity $\iff \pi_1(\mathcal{X})'' \simeq \pi_2(\mathcal{X})'' \iff c(\pi_1) = c(\pi_2) \iff W^{*}(\pi_1)_{*} = W^{*}(\pi_2)_{*}$)

***Quasi-equivalence and Modular Structure***

(iii) Representation $(\pi^{\overset{|\,|}{\circ\circ}}, c(\pi)\mathfrak{H}_{u})$ of the von Neumann algebra $W^{*}(\pi) \simeq \pi^{\overset{|\,|}{\circ\circ}}(\mathcal{X})''$ in $c(\pi)\mathfrak{H}_{u} = P(\pi^{\overset{|\,|}{\circ\circ}})\mathfrak{H}_{u}$ gives the ***standard form*** of $W^{*}(\pi)$ equipped with a normal faithful semifinite weight $\varphi$ and the associated Tomita–Takesaki ***modular structure*** $(J_{\varphi}, \Delta_{\varphi})$ [5], whose universality is characterized by the adjunction,

$$Std(\pi^{\overset{|\,|}{\circ\circ}} \to \sigma) \simeq Rep_{\mathcal{X}}(\pi \to \sigma).$$

Namely, any intertwiner $T \in Rep_{\mathcal{X}}(\pi \to \sigma)$ to a standard form representation $(\sigma, \mathfrak{H}_{\sigma})$ of $W^{*}(\sigma)$ is uniquely factored $T = T^{\overset{|\,|}{\circ\circ}} \circ \eta_{\pi}$ through the canonical homotopy $\eta_{\pi} \in Rep_{\mathcal{X}}(\pi \to \pi^{\overset{|\,|}{\circ\circ}})$ with $\exists! T^{\overset{|\,|}{\circ\circ}} \in Rep_{\mathcal{X}}(\pi^{\overset{|\,|}{\circ\circ}} \to \sigma)$.

***Quasi-equivalence and Sector-classifying Groupoid***

(iv) Quasi-equivalence $\pi_1 \approx \pi_2$ defines the sector-classifying groupoid $\Gamma_{\approx}$ consisting of ***invertible intertwiners*** in $Rep_{\mathcal{X}}$, which reduces on each $\pi \in Rep_{\mathcal{X}}$ to the automorphism group, $\Gamma_{\approx}(\pi, \pi) = Aut(W^{*}(\pi)) \simeq Isom(W^{*}(\pi)_{*})$, isomorphic to the isometry group of predual $W^{*}(\pi)_{*}$.

Modular structure (iii) of a von Neumann algebra $W^{*}(\pi) := \mathcal{M}$ in the standard form $(\pi^{\overset{|\,|}{\circ\circ}}, c(\pi)\mathfrak{H}_{u})$ can be understood as the ***unitary implementation*** of a normal subgroup $G_{\mathcal{M}} := Isom(\mathcal{M}_{*})^{\mathcal{M}} \lhd Isom(\mathcal{M}_{*})$ fixing $\mathcal{M}$ pointwise by the unitary group $\mathcal{U}(\mathcal{M}')$ of the commutant $\mathcal{M}'$: namely, for $\gamma \in G_{\mathcal{M}}$, $\exists U'_{\gamma} \in \mathcal{U}(\mathcal{M}')$ such that $\langle \gamma\omega, x \rangle = \langle \omega, \gamma^{*}(x) \rangle = \langle \omega, U'^{*}_{\gamma} x U'_{\gamma} \rangle$ for $\omega \in \mathcal{M}_{*}$, and $U'^{*}_{\gamma} x U'_{\gamma} = x \iff x \in \mathcal{M}$. Through the modular conjugation $J_{\varphi}(-)J_{\varphi}$, this unitary group can naturally be related to the modular group $\Delta_{\varphi}^{it}$.

***Quasi-equivalence and Galois Structure***

(v) If $\mathcal{M}$ is of type III, the following Galois-type relations hold involving the crossed product by a coaction of $G_{\mathcal{M}} \simeq \mathcal{U}(\mathcal{M}')$ on $\mathcal{M}$:

$$\mathfrak{Z}(\mathcal{M})' = \mathcal{M} \vee \mathcal{M}' = \mathcal{M} \rtimes \widehat{G_{\mathcal{M}}} : \text{Galois extension of } \mathcal{M},$$
$$\mathcal{M} = (\mathcal{M} \vee \mathcal{M}')^{G_{\mathcal{M}}} : \text{fixed-point subalgebra under } G_{\mathcal{M}},$$
$$G_{\mathcal{M}} = Gal(\mathfrak{Z}(\mathcal{M})'/\mathcal{M}) : \text{Galois group of } \mathcal{M} \hookrightarrow \mathfrak{Z}(\mathcal{M})',$$

according to which trivial center $\mathfrak{Z}(\mathcal{M}) = \mathbb{C}\mathbf{1}$ to characterize a ***sector*** can be reinterpreted as ***ergodicity*** condition on $\mathcal{M}$ under $Aut(\mathcal{M})$ or $G_{\mathcal{M}}$:

$$\mathbb{C}\mathbf{1} = \mathcal{M} \cap \mathcal{M}' = \mathcal{M}' \cap \mathcal{U}(\mathcal{M}')' = (\mathcal{M}')^{G_{\mathcal{M}}} \supset (\mathcal{M}')^{Aut(\mathcal{M})}.$$

## 4.7 Symmetry Breaking Creates Symmetric Space

When a domain of physical nature is formulated theoretically according to our quadrality scheme, we see in the following that a homogeneous space $G/H$ emerges in the sector classifying space if the symmetry of the system described by a group $G$ is broken up to its subgroup $H$ remaining unbroken. In this situation, $G/H$ is shown to be a ***symmetric space*** [10] (which satisfies various nice properties as shown below), according to the definition and the criterion of symmetry breaking. For this reason, the induced representation $Ind_H^G$ [7] to describe the broken symmetry $G$ on the sector classifying space $G/H$ as a symmetric space has inevitably a strong connection with automorphic forms and the zeta functions playing important roles in number theory.

The mutual relation between the quadrality scheme and the groups to describe symmetries can be depicted as follows:

| | ***Spec=*** ***classifying space*** | = | $G/H$ | : Visible Macro |
|---|---|---|---|---|
| Emergence ↗ | ⇅ | | ↑ | |
| ***(Family of) States*** | ⇆ ***Algebra*** | ↶ | $G$ | : Micro-Macro boundaries : ***t*-ch** |
| | ⇅ ↗ | | ↑ | |
| | ***Dynamics*** | ↶ | $H$ | : Invisible Micro: s-**ch** |

### 4.7.1 Symmetry Breaking

According to [1] the definition of a symmetry breaking can be given as follows:

**Definition** (***Symmetry Breaking***): Let $\mathcal{X}$ be a C*-algebra describing quantum fields with an automorphic action $\tau{:}\mathcal{X} \underset{\tau}{\curvearrowleft} G$ of a Lie group $G$ on $\mathcal{X}$ and $(\pi, \mathfrak{H})$ be a representation of $\mathcal{X}$. If the spectrum $Spec(\mathfrak{Z}_\pi(\mathcal{X}))$ of its center $\mathfrak{Z}_\pi(\mathcal{X}) = \mathfrak{Z}(\pi(\mathcal{X})'')$ is pointwise $G$-invariant (almost everywhere w.r.t. the central measure), the symmetry $(G, \tau)$ on $\mathcal{X}$ is said to be ***unbroken*** in $(\pi, \mathfrak{H})$ and ***broken*** otherwise.

The reason for the complicated situations concerning the symmetry breaking in quantum field theory is due to such a contrast between the quantum systems with *finite* versus *infinite* degrees of freedom: while the use of a unitary representation $U$ of $G$ leads automatically to the ***unbroken*** symmetry (which is always the case for systems with *finite* degrees of freedom), the very non-existence of $U$ realizable only in those with *infinite* degrees of freedom characterizes the ***broken*** symmetry. This is the reason why we need $G$-actions both in ***C*****-** and ***W*-versions*** in the above criterion for symmetry breaking. In order to streamline the discussion, we introduce an "***augmented algebra***" [1] defined by a C*-crossed product[1] $\mathcal{X} \rtimes \widehat{(H\backslash G)} =: \widehat{\mathcal{X}}$ of $\mathcal{X}$ together with the dual $\widehat{(H\backslash G)}$ of $(G/H)$ with $H$ an unbroken subgroup of $G$, which allows the unitary implementation of the broken $G$ at the expense of the appearance of a non-trivial center $\mathfrak{Z}_\pi(\widehat{\mathcal{X}})$ with $Spec(\mathfrak{Z}_\pi(\widehat{\mathcal{X}})) = G/H$ in the representation of $\widehat{\mathcal{X}}$. In this way, the corresponding von Neumann algebra $\pi(\mathcal{X})''$ can be taken as $(\pi \rtimes U_\tau)(\widehat{\mathcal{X}})''$ in the latter half of the above definition.

As will be shown later in Sect. 4.11, the appearance of a central spectrum as $Spec(\mathfrak{Z}_\pi(\widehat{\mathcal{X}})) = G/H$ suggests an interesting commitment of ***induced representations*** and ***imprimitivity*** [7] involving the following exact sequences:

$$\mathrm{Rep}(G/H) \hookrightarrow \mathrm{Rep}(G) \overset{Ind_H^G}{\leftrightarrows} \mathrm{Rep}(H),$$

$$H \underset{(Ind_H^G)^*}{\rightleftarrows} G \twoheadrightarrow G/H.$$

Here the bigger group $G$ can be viewed as a principal $H$-bundle with a structure group $H$ over the base space $G/H = Spec(\mathfrak{Z}_\pi(\widehat{\mathcal{X}}))$ as the sector classifying space, and the dual map $(Ind_H^G)^*$ of $Ind_H^G$ (sometimes called "Wigner rotation") plays the role of a gauge connection.

We note here that the starting point in our present discussion is just a C*-dynamical system $\mathcal{X} \underset{\tau}{\curvearrowleft} G$ given by the C*-algebra $\mathcal{X}$ of quantum fields acted upon by a Lie group $G$ of the symmetry of the system. In a sense, however, the spacetime background of the dynamical system $\mathcal{X} \underset{\tau}{\curvearrowleft} G$ without being mentioned at the beginning, has emerged automatically in the form of $G/H$ as a result of the symmetry breaking from $G$ to $H$. In this sense, the essence of symmetry breaking is crucial and universal for general understanding of the contents and of the meaning of the above quadrality scheme for Micro–Macro duality.

For this purpose, we remark first such a crucial point that the center of the represented algebra $\pi(\widehat{\mathcal{X}})''$ consisting of the Macro variables of the system as the low energy modes has such a spectrum as $Spec(\mathfrak{Z}_\pi(\widehat{\mathcal{X}})) = G/H$ whose non-trivial motion is driven by the action of $G$ to exhibit the essence of symmetry breaking as the "infrared instability". Arbitrary representations of $\widehat{\mathcal{X}}$ are decomposed into the

[1]While the contents of [9] is the crossed products of von Neumann algebras, this book is very instructive for understanding the essence of the concept, either in C*- or W*-versions.

direct sum of $G$-*unbroken* factor representations and $G$-centrally ergodic non-factor representations (the latter ones corresponding to symmetry breaking) according to which a "phase diagram" can be drawn on the central spectrum.

While the vivid motions of nature cannot be grasped without stepping into the microscopic dynamics, a microscopic process itself is invisible as such without any devices to visualize and/or magnify it. Then we can naturally understand that how unrealistic it is to assume that consistent description of a phenomenon is possible anytime anywhere. If so, we cannot implement a logically consistent description of a phenomenon without solving the "inverse problem" of retrospective inferences, starting from such macroscopic data obtainable at the stabilized Macro level, to aim at identifying the dynamical Micro-system which has caused the phenomenon in question.

### *4.7.2 Symmetry Breaking and Symmetric Spaces*

The breakdown of a symmetry described by a Lie group $G$ with the associated Lie algebra $\mathfrak{g}$ creates an interesting Micro–Macro interface between the Micro level invariant under the unbroken Lie subgroup $H$ with its Lie algebra $\mathfrak{h}$ and the visible Macro level of the sector classifying space $M = G/H$ formed in the emergence of condensed order parameters which parametrize the so-called "degenerate vacua" arising from the symmetry breaking. Namely, the criterion for the symmetry breaking ensures that $M = G/H$ becomes a ***symmetric space*** (É. Cartan) [10] whose Lie structure $\mathfrak{m} = \mathfrak{g}/\mathfrak{h}$ is characterized locally by the relation $[\mathfrak{m}, \mathfrak{m}] \subset \mathfrak{h}$ [11].

Here the commutator set $[\mathfrak{m}, \mathfrak{m}]$ of tangent vectors in $M$ describes the holonomy effect of the curvature of $M$ in the loop motions on $M$. Since a trajectory forming a loop returns to its starting point on the sector classifying space $M$, the net effect of the loop reduces to such components of the transformation group as fixing the sector unchanged, being contained in the unbroken symmetry corresponding to $\mathfrak{h}$ which can be expressed as macroscopic loops $[\mathfrak{m}, \mathfrak{m}]$ penetrated by microscopic arrows in $\mathfrak{h}$. We note that the relation $[\mathfrak{m}, \mathfrak{m}] \subset \mathfrak{h}$ imposes such a restriction on the possible symmetry breaking pattern that the unbroken symmetry can be abelian $\mathfrak{h} = 0$ only if the broken symmetry is abealin: $\mathfrak{m} = 0$.

### *4.7.3 Examples of Symmetric Spaces: Chiral Symmetry and Lorentz Boosts*

***Chiral symmetry*** in "current algebra": the typical examples of the symmetry breaking yielding symmetric space structures can be found in the chiral symmetry of current algebra commutation relations:

$$[V, V] = V, [V, A] = A, [A, A] = V,$$

$$(V \in \mathfrak{h} : \text{vector currents}, A \in \mathfrak{m} : \text{axial currents}).$$

Choosing the Lorentz group $L_+^\uparrow$ as $G$ and the rotation group $SO(3)$ as the unbroken subgroup $H$ of $G$, we can find a symmetric space $M = G/H \cong \mathbb{R}^3$ given as the space of all the Lorentz frames mutually connected by the Lorentz boosts.

In fact, with definitions $\mathfrak{h} := \{M_{ij}; i, j = 1, 2, 3, i < j\}, \mathfrak{m} := \{M_{0i}; i = 1, 2, 3\}$, we can extract the relations $[\mathfrak{h}, \mathfrak{h}] = \mathfrak{h}$, $[\mathfrak{h}, \mathfrak{m}] = \mathfrak{m}$, $[\mathfrak{m}, \mathfrak{m}] \subset \mathfrak{h}$ from the following Lie algebra structure of the Lorentz group:

$$[iM_{\mu\nu}, iM_{\rho_\sigma}] = -(\eta_{\nu\rho} iM_{\mu\sigma} - \eta_{\nu\sigma} iM_{\mu\rho} - \eta_{\mu\rho} iM_{\nu\sigma} + \eta_{\mu\sigma} iM_{\nu\rho}).$$

While both the breaking of rotation group and of Lorentz boosts are rarely recognized, the cases with ***Lorentz boosts*** unbroken is actually exceptional aside from the vacuum situation [$\cdot$.$\cdot$] the reason for this can be understood with the help of Borchers-Arveson-Araki theorem [12] in such a way that the energy as a physical quantity can be observable only in the vacuum representation with $T = 0K$ and that Lorentz boosts correspond to the broken symmetry for $T \neq 0$ [13]. In this sense, the totality $M = L_+^\uparrow/SO(3) \cong \mathbb{R}^3$ of Lorentz frames with the broken Lorentz symmetry provides the typical example of symmetry breaking causing the emergence of symmetric space characterized by the relation [boost, boost]= rotation.

### *4.7.4 Second Law of Thermodynamics as an Example of Symmetric Space*

The essence of the characterization of symmetric spaces by "macroscopic loops $[\mathfrak{m}, \mathfrak{m}]$ penetrated by microscopic arrows in $\mathfrak{h}$" can be exhibited directly in the macroscopic world in the form of the second law of thermodynamics. Its mathematical essense can be seen in the following exact sequence[2]

$$\Delta' Q \overset{q}{\hookrightarrow} \Delta E = \Delta' Q + \Delta' W \overset{p}{\twoheadrightarrow} \Delta' W,$$

$$\text{i.e., } \mathrm{Im}(q) = \ker(p),$$

This is the same as the relation $\mathfrak{h} \hookrightarrow \mathfrak{g} \twoheadrightarrow \mathfrak{m} = \mathfrak{g}/\mathfrak{h}$ to characterize the Lie structure of the homogeneous space $M = G/H$.

---

[2]Equality $\mathrm{Im}(q) = \ker(p)$ means that the vanishing energy balance ($\ker(p)$) taken as the visible work is equivalent to the input-output of the heat ($\mathrm{Im}(q)$).

The cyclic processes of a heat engine corresponds to the loops on the thermodynamic phase space $M$ described by thermodynamic variables and the holonomy $[\mathfrak{m}, \mathfrak{m}]$ associated with such cycles describes the incoming and outgoing heat between the heat engine and the external world in combination with the relations $[\mathfrak{m}, \mathfrak{m}] \subset \mathfrak{h}$ and $\Delta E = \Delta' Q + \Delta' W = 0$: $-\Delta' W = -[\mathfrak{m}, \mathfrak{m}] = \Delta' Q > 0$, in which the characterization of $M$ as a symmetric space $[\mathfrak{m}, \mathfrak{m}] \subset \mathfrak{h}$ corresponds to the ***second law of thermodynamics*** in ***Kelvin's version!***

## 4.8 Sector Bundle and Holonomy

In the case of symmetry breaking of $G$ up to its unbroken compact subgroup $H$, the sector structure should be understood in two levels, one with $\widehat{H}$, the dual of unbroken $H$ consisting of its irreducible representations, and the other with $G/H$ corresponding to the broken $G$. To unify these two levels, it is convenient to introduce the concept of a ***sector bundle***:

$$\widehat{H} \hookrightarrow G \underset{H}{\times} \hat{H} \twoheadrightarrow G/H.$$

In this context we can see the physical origin of space-time concept in its ***physical emergence process*** [14].

For simplicity, we assume here that a group $G$ of broken internal symmetry be extended by a group $\mathcal{R}$ of space-time symmetry (typically translations) into a larger group $\Gamma = \mathcal{R} \dot{\times} G$ defined by a semi-direct product of $\mathcal{R}$ and $G$ with $\Gamma/G = \mathcal{R}$.

In this case, the sector bundles have a double fibration structure:

$$\begin{array}{ccccc} \widehat{H} \hookrightarrow & G \underset{H}{\times} \hat{H} & \hookrightarrow & \Gamma \underset{G}{\times} (G \underset{H}{\times} \hat{H}) = \Gamma \underset{H}{\times} \hat{H} & \\ & \downarrow & & \downarrow & . \\ & G/H & & \Gamma/G = \mathcal{R} & \end{array}$$

### 4.8.1 Holonomy along Goldstone Condensates

Thus, we see that Spec= sector-classifying space has three different axes on different levels:

(i) sectors $\widehat{H}$ of ***unbroken*** symmetry $H$,

(ii) degenerate vacua $G/H = M$ due to ***broken internal*** symmetry [1, 15],

(iii) $\Gamma/G = \mathcal{R}$ as emergent ***space-time*** [14] in broken external symmetry.

These axes appear geometrically as a series of structure group contractions $H \leftarrow G \leftarrow \Gamma$ of principal bundles $P_H \hookrightarrow P_G \hookrightarrow P_\Gamma$ over $\mathcal{R}$, specified by ***solderings*** as bundle sections, $\mathcal{R} \overset{\rho}{\hookrightarrow} P_G/H = P_H \underset{H}{\times} (G/H), \mathcal{R} \overset{\tau}{\hookrightarrow} P_\Gamma/G = P_G \underset{G}{\times} (\Gamma/G) = P_G \underset{G}{\times} \mathcal{R}$, which correspond physically to ***Goldstone modes***:

$$\begin{array}{ccccc}
P_H & \hookrightarrow & P_G & \hookrightarrow & P_\Gamma \\
H\downarrow & \circlearrowleft & \downarrow H & \circlearrowleft & \downarrow H \\
\mathcal{R} & \overset{\rho}{\hookrightarrow} & P_G/H & \overset{\sigma}{\hookrightarrow} & P_\Gamma/H \\
 & \diagdown\diagdown \circlearrowleft & \downarrow G/H & \circlearrowleft & \downarrow G/H \\
 & & \mathcal{R} & \overset{\tau}{\hookrightarrow} & P_\Gamma/G \\
 & & & \diagdown\diagdown \circlearrowleft & \downarrow \mathcal{R} \\
 & & & & \mathcal{R}
\end{array}$$

### *4.8.2 Helgason Duality with Hecke Algebra*

From the algebraic viewpoint (dual to ***Helgason duality*** [16] $K\backslash G \leftrightarrow G/H$:

$$\begin{array}{ccccc}
 & \nearrow & K\backslash G/H & \nwarrow & \\
K\backslash G & & \leftrightarrow & & G/H \\
 & \nwarrow & G & \nearrow &
\end{array}$$

with Radon transform and ***Hecke algebra*** $K\backslash G/H$), the essence of the relevant structures can be viewed as the "stereo-graphic" extension of planar diagrams controlling "***augmented algebras***" [1] of crossed products to describe symmetry breaking:

$$\begin{array}{ccccc}
{}_{G/H}\swarrow & \mathcal{X}^H = \widetilde{\mathcal{X}}^G & \searrow_H & \\
\widetilde{\mathcal{X}}^H & \Downarrow & & \mathcal{X} \\
\downarrow_H \searrow\searrow & \Downarrow & \swarrow_{G/H} & \downarrow \\
\downarrow & \widetilde{\widetilde{\mathcal{X}}} & & \downarrow \\
\downarrow \quad \swarrow & \Downarrow & \searrow\searrow & \downarrow \\
\widehat{H\backslash G} & \hookrightarrow \widehat{G} & \twoheadrightarrow & \widehat{H}
\end{array}
\quad \rightleftarrows \quad
\begin{array}{ccccc}
{}_{\mathcal{R}}\swarrow & \mathcal{O}_\rho = \mathcal{O}_d^H & \searrow_H & \\
\mathcal{A}(\mathcal{R}) & \Downarrow & & \mathcal{O}_d \\
\downarrow_H & \searrow\searrow\Downarrow & \swarrow_{\mathcal{R}} & \downarrow \\
\downarrow & \mathcal{X}(\mathcal{R}) & & \downarrow \\
\downarrow \quad \swarrow & \Downarrow & \searrow\searrow & \downarrow \\
\widehat{\mathcal{R}} & \hookrightarrow \widehat{\Gamma} & \twoheadrightarrow & \widehat{H}
\end{array} .$$

[same sort of lines constitute exact sequences]

Note that the push-out diagram appearing in the Doplicher-Roberts reconstruction [17] of field algebra $\mathcal{X}(\mathcal{R})$ shows up here (on the right) in spite of its unbroken symmetry.

### 4.8.3 Symmetric Space Structure= Maxwell-Type Equation as a Result of Symmetry Breaking

The symmetric space structures of $G/H = M$ and $\Gamma/G = \mathcal{R}$ arising from the symmetry breaking are characterized by the equations of type $[\mathfrak{m}, \mathfrak{m}] \subset \mathfrak{h}$, which connects the holonomy $[\mathfrak{m}, \mathfrak{m}]$ (in terms of curvature) with generators $\mathfrak{h}$ of unbroken subgroup.

What is really interesting here is to note that this feature is shared in common by the Maxwell and Einstein equations of electromagnetism and of gravity, respectively:

LHS: (curvature $F_{\mu\nu}$ or $R_{\mu\nu}$) = (source current $J_{\mu}$ or $T_{\mu\nu}$) : RHS, if one realizes that all the quantities on the left-hand side, $[\mathfrak{m}, \mathfrak{m}]$, $F_{\mu\nu}$ and $R_{\mu\nu}$ represent the holonomy terms and that those on the right-hand side are associated with generators $\mathfrak{h}$ of unbroken subgroups.

According to the standard viewpoint, Maxwell equation is an identity following from the invariance of "action integral" under space-time dependent transformations in the context of the second Noether theorem (developed in the theory of invariants). In contrast, *no such classical quantities as action integrals nor Lagrangian densities* are defined in our algebraic and categorical formulation of quantum fields.

In spite of the absence of such quantities, symmetry breaking criterion with $[\mathfrak{m}, \mathfrak{m}] \subset \mathfrak{h}$ tells us that Maxwell-type equation with curvature term $[\mathfrak{m}, \mathfrak{m}]$ on the left-hand side and the internal symmetry term $\mathfrak{h}$ on the right-hand side is just a consequence of symmetry breaking of local gauge invariance into spacetime and internal symmetries. Putting the Clebsch-dual electromagnetic field $S_{\mu\nu}$ [18] in the place of $[\mathfrak{m}, \mathfrak{m}]$, therefore we can learn that $S_{\mu\nu}$ represents the condensation effect of dressed photons.

### 4.8.4 Galois Functor in Doplicher-Roberts Reconstruction of Symmetry

The expected roles of action integral in the usual contexts is to determine representation contents of a theory. In view of the essential difficulty in making mathematical sense from action interal, we replace it with categorical data concerning Galois group due to Doplicher and Roberts (DR) [17], based on DR category $\mathcal{T}$ of modules of local excitations: objects of $\mathcal{T}$ are local *endomorphisms* $\rho \in End(\mathcal{A})$ of observable algebra $\mathcal{A}$, selected by DHR localization criterion [19] $\pi_0 \circ \rho \restriction_{\mathcal{A}(\mathcal{O}')} \cong \pi_0 \restriction_{\mathcal{A}(\mathcal{O}')}$, and morphisms are $T \in \mathcal{T}(\rho \to \sigma) \subset \mathcal{A}$ intertwining $\rho, \sigma \in \mathcal{T}$: $T\rho(A) = \sigma(A)T$.

The group $H$ of unbroken internal symmetry arises as the group $H = End_{\otimes}(V)$ of unitary tensorial (=monoidal) natural transformations $u : V \to V$ with the representation functor $V : \mathcal{T} \hookrightarrow Hilb$ to embed $\mathcal{T}$ into the Hilbert-space category $Hilb$ with morphisms as bounded linear maps.

### 4.8.5 Galois Functor in Category and Its Local Gauge Invariance

Recall that a natural transformation $u : V \to V$ is characterized by the commutativity diagrams: $\begin{array}{ccc} V(\rho) & \overset{u_\rho}{\to} & V(\rho) \\ V(T) \downarrow & \circlearrowleft & \downarrow V(T) \\ V(\sigma) & \underset{u_\sigma}{\to} & V(\sigma) \end{array}$, namely, $V(T)u_\rho = u_\sigma V(T)$ for $T \in \mathcal{T}(\rho \to \sigma)$.

Our simple proposal here is to define a *local gauge transformation* $\tau_u(V)$ of functor $V$ by $\tau_u(V)(T) := u_\sigma V(T)u_\rho^{-1}$ corresponding to a natural transformation $u \in H = End_\otimes(V)$.

Then, the above equality, $u_\sigma V(T) = V(T)u_\rho$, can be reinterpreted as *local gauge invariance* $\tau_u(V) = V$ of functor $V$ under *local gauge transformation* $V \to \tau_u(V)$ induced by a natural transformation $u \in H = End_\otimes(V)$, as has been visualized in the context of lattice gauge theory.

### 4.8.6 Local Gauge Invariance and Maxwell Equation

In the original Doplicher-Roberts theory [17], local $endomorphismsas$ $\rho \in \mathcal{T} \subset End(\mathcal{A})$ have, unfortunately, been regarded as *global* constant objects, owing to the emphasis on space-time transportability,[3] and hence, the left-right difference of $u_\rho$ and $u_\sigma$ in $\tau_u(V)(T) := u_\sigma V(T)u_\rho^{-1}$ has not been properly recognized as important signal of local gauge structures.

From the general viewpoint of forcing method, however, the essential features of logical extension ***from constants to variables*** naturally lead to the interpretation of $\tau_u(V)(T) = u_\sigma V(T)u_\rho^{-1} = V(T)$ as the characterization of local gauge invariance of $V$ under local gauge transformation $u : \mathcal{T} \ni \rho \longmapsto u_\rho$.

This is in harmony also with the alternative formulation of principal bundles in terms of group-valued Čech cohomologies.

### 4.8.7 Symmetry Breaking and Maxwell Equation

In the above proto-type discussion of the DR reconstruction, the recovered group $H$ of unbroken symmetry is compact. So, the space $\hat{H}$ of sector parameters has discrete structure, which makes it difficult to derive differential equations. Noting that the essential roles played by DR category $\mathcal{T} \subset End(\mathcal{A}) = End(\mathcal{X}^H)$ are to determine the factor spectrum $Sp(\mathfrak{Z}(\mathcal{X}^H)) = \hat{H}$, we replace $\mathcal{T}$ by $\approx{\mathcal{T}} = End(\approx{\mathcal{X}}^H)$ with $\approx{\mathcal{X}} = \mathcal{X}^H \rtimes \widehat{\mathcal{R}}$ and with $\Gamma/G = \mathcal{R}$ (: spacetime) concerning the two-step construction

[3] This has led to the mathematical definition of "sectors" of $\mathcal{A}$ by $End(\mathcal{A})/Inn(\mathcal{A})$.

of augmented algebras associated with the series of group contractions/extensions: $H \hookrightarrow G \hookrightarrow \Gamma$. By repeating the categorical formulation of $End_{\otimes}(V : \mathcal{T} \hookrightarrow Hilb)$ with $\mathcal{T}$ and $V$ replaced by $\tilde{\tilde{\mathcal{T}}}$ and $\tilde{\tilde{V}}$, we can reproduce the essence of second Noether theorem to connect the local gauge invariance and Maxwell equation.

Already in the previous section discussing the symmetric space structure inherent in Maxwell equation, we draw the conclusion of relating symmetry breaking and Maxwell equation in the case of Clebsch-dual field appearing in the dressed photon phenomenon.

## 4.9 Symmetry Breaking, Saddle Points and Indefinite Metric in Emergence to Macro

To construct a consistent theory for describing dressed photons, it will be a crucial point to recognize the proper dynamic functions in close relation with the "tapering" cone structure formed by the condensed dressed photons. To implement the ideas in this direction, it is important to install the ***Clebsch-dual variables*** [18] discovered by Sakuma which carry ***spacelike momenta*** and constitute the characteristic ***off-shell structure*** of electromagnetic field.

In this context related with the understanding of condensed order parameters, we compare first the Bardeen-Cooper-Schrieffer (BCS) theory of superconductivity and the relativistic formulation of the Higgs mechanism from the above viewpoint, both of which are usually understood as the similar theories for accommodating massive gauge bosons without contradicting the local gauge invariance of the gauge theory. Between these two formulations we notice such important differences as the existence of Cooper pair mode in the former and its absence in the latter, for instance. Actually, what plays crucial roles at this point is just the effects related with the ***indefinite metric***, since the elimination of the former can be traced to the violation of the subsidiary condition which guarantees the positivity [20].

When we try to re-examine the roles played by the "indefinite metric" in the wider contexts including thermodynamics and statistical mechanics, gauge theories and induced representations of groups, the important aspects common in various cases can be found in the existence of ***saddle points*** and the roles played by "***indefinite metric***" at those saddle points which is due to the co-existence of ***stable and unstable*** directions.

When this mechanism for determining geometric invariants is applied to the sector classifying space, the non-trivial relations between the dynamics of Micro quantum system and the geometric structure of classifying space at the Macro level can be envisaged and described in terms of the ***unstable modes*** and the indefinite metric corresponding to the saddle point structures. In the Morse-theoretical contexts [21] of deriving homologies and/or cohomologies as geometric invariants, they are determined by the negative-metric components of the ***Hessians*** defined as the second derivatives of ***Morse functions*** on the manifold, and hence the dimensionality of this negative metric is called "***Morse index***".

If concrete systematic descriptions of dynamical processes become possible from this viewpoint, the actual meaning of treating "stability" aspects would be reduced to examining such problems as which "branches" would satisfy the (conditional) stability and which conditions can support the classifying space *Spec* describing the multi-sector structure serves as the setting up for such discussions. In this way, would it be possible for us to envisage the problems of whether stable or unstable naturally in wide perspectives? Moreover, this kind of contexts would require us to pursue such processes as the formation of classifying spaces *Spec* through the emergences triggered by the instability at the saddle points as the bifurcation points between stability and instability. Through this kind of changes, big transitions would perhaps be implemented to enable us to be faithful to such natural recognition that ***dynamical motions are absolute and fundamental*** and ***stable states are conditional***.

Are the basic points for this direction hidden in the "indefinite metric" which has been disliked so far?: the answer to this question is really affirmative when we combine the following three points, (i) ***indefinite metric at the saddle point***, (ii) ***symmetry breaking aspects inherent in Maxwell equation***, and (iii) the ***spacelike supports of dressed photon momenta described by Clebsch-dual field***.

## 4.10 Roles Separated into Micro Versus Macro and Geometric Invariants

Now we consider the problems according to the above explanations. For this purpose, we consider first (1) induced representation of groups, and (2) guage theories.

(1) Whether a Lie group $G$ is compact or not is determined by the Killing form $\theta$ of the associated Lie algebra $\mathfrak{g}$ in such a way that negative definiteness of $\theta$ corresponds to the compactness of $G$ and that indefinite $\theta$ to the non-compact $G$.

While the irreducible representation $(\sigma, W)$ of the maximally compact subgroup $H$ is realized in a (finite-dimensional) positive definite Hilbert space $W$, the irreducible finite-dimensional representation of the non-compact semisimple $G$ is possible only in a vector space ***with indefinite metric***. In this situation, the induced representation $Ind_H^G(\sigma)$ [7] of $G$ induced from a representation $(\sigma, W)$ of $H$ can be realized in an *infinite-dimensional* positive definite Hilbert space $L^2_\sigma(G \to W) = L^2(G) \underset{H}{\otimes} W$ which is defined as the subspace of $W$-valued functions $\xi : G \longrightarrow W$ on $G$ satisfying the condition (of $H$-equivariance):

$$\xi(gh) = \sigma(h^{-1})\xi(g) \text{ for } g \in G \text{ and } h \in H.$$

According to the equivariance condition, the representation $(\sigma, W)$ of $H$ is recovered (by the left translation) at the origin $e \in G$: $[l_h\xi](e) = \xi(h^{-1}e) = \xi(eh^{-1}) = \sigma(h)\xi(e)$.

In this way the appearance of indefinite metric in the representation space due to non-compactness of $G$ is resolved into the infinite dimensionality of the representation space.

(2) In the case of (abelian) gauge theory with a gauge potential $A_\mu$, its Lorentz covariant formulation is possible only in a state vector space with an indefinite metric. In the total space with indefinite metric, we can introduce the concept of a physical subspace $\mathcal{V}_{phys}$ consisting of gauge-invariant physical modes, by imposing such a "subsidiary condition" [20] as $\Phi \in \mathcal{V}_{phys} \Longleftrightarrow (\partial_\mu A^\mu)^{(+)}\Phi = 0$. In this physical subspace $\mathcal{V}_{phys}$ longitudinal modes causing the difficulties of indefinite metric are shown to be absent, according to which consistency of the probabilistic interpretation is guaranteed within $\mathcal{V}_{phys}$ at the microscopic level.

Existence of quantum modes with "indefinite metrics" spoils the consistency of the theory at Micro levels, as is seen in the difficulties caused by the longitudinal photons on the probabilistic interpretation. For this reason, one tries to exclude longitudinal photons from QED and it is the common wisdom that such unphysical modes can be systematically expelled from the physical subspace of physical modes selected by imposing a suitable "subsidiary condition".

If we look at the macroscopic real world without prejudice, however, it is a plain fact that there exist Coulomb modes which describe the interaction potentials between electric charges. According to the *standard* "quantum-classical correspondence", the mutual relations between Micro and Macro, between quantum and classical, can be understood in such a way that the quantum observables non-commutative in the microscopic scales become mutually commutative classical observables in the "classical limit" with $\hbar \to 0$ and that the classical observables can be "quantized" through imposing the canonical commutation relations as a result of which quantum theory equipped with non-commutative quantum observables can be realized.

In such cases that emergence processes into Macro plays essential roles, however, the above simple-minded picture between quantum and classical observables cannot be maintained, where we encounter such paradoxical situations that some physical variables invisible (or driven away as unphysical modes) at Micro level may become visible in the Macro world, as is exemplified by the longitudinal Coulomb modes. In such cases, how about the fate of "indefinite metric" which is to be viewed as risky?

## 4.11 Alternation between Indefinite Metric in Micro and Geometric Non-triviality in Macro

What looks mysterious is the disappearance of indefinite metric in macroization emergence and geometric non-triviality as its substitute which takes place in the construction of representations of non-compact groups induced from its compact subgroup.

Typical example can be found in the infinite-dimensional unitary representation of the (inhomogeneous) Lorentz group $(\mathbb{R}^4 \rtimes)SL(2, \mathbb{C})$, which was first established in 1939 by a physicist E. Wigner [22] in use of the method of induced representations. While a non-compact group appears in the form of $SL(2,\mathbb{C})$, we do not encounter the indefinite metric either in this situation.

The mechanism of induced representations to suppress the indefinite metric can be seen in such a form that the non-compact group $SL(2,\mathbb{C})$ possibly inducing the indefinite metric is treated here as the base space $M := G/H = SL(2,\mathbb{C})/SU(2)$ of the $SU(2)$-bundle:

$$H := SU(2) \hookrightarrow G = SL(2,\mathbb{C}) \twoheadrightarrow M = SL(2,\mathbb{C})/SU(2),$$

and, at each point of this base space $M$ (which is a part of sector classifying space), we have a fixed Lorentz frame acted upon by the rotation group $SU(2)$ as the structure group of each Lorentz frame and the actions of the Lorentz boosts $SL(2,\mathbb{C})$ are just to move from one Lorentz frame to another, which do not exhibit the indefinite metric related with $SL(2,\mathbb{C})$ like the case of its matrix representation. On this geometric setting up, the representation $Ind_{SU(2)}^{SL(2,\mathbb{C})}(\sigma) \in \mathrm{Rep}(G)$ induced from a representation $\sigma \in \mathrm{Rep}(H)$ is defined on the Hilbert space $L^2_\sigma(G \to W)$ as given above, which is isomorphic to $L^2(G/H) \otimes W$ in the present situation where the base space $M = SL(2,\mathbb{C})/SU(2)$ is a symmetric space.

In view of the duality,

$$[Ind_H^G(\sigma)](g) = \langle g | Ind_H^G(\sigma)\rangle = \langle (Ind_H^G)^*(g)|\sigma\rangle = \sigma((Ind_H^G)^*(g)),$$

we note that each group element $g \in G$ belonging to the non-compact group $G = SL(2,\mathbb{C})$ is transferred to $(Ind_H^G)^*(g)$ belonging to the compact subgroup $H := SU(2)$:

$$\mathrm{Rep}(SU(2)) \ni \sigma \longmapsto Ind_H^G(\sigma) \in \mathrm{Rep}(SL(2,\mathbb{C})),$$

$$SU(2) \ni (Ind_H^G)^*(g) \leftarrow g \in SL(2,\mathbb{C}).$$

This mapping $(Ind_H^G)^*$ is called (in physics) "Wigner rotation", since each of its image $(Ind_H^G)^*(g) \in SU(2)$ is a rotation.

According to the exact sequence $H \hookrightarrow G \twoheadrightarrow M = G/H$, we note that the group $G$ can be interpreted as an $H$-principal bundle over the base space $M = G/H$ with $H$ as its structure group. In this context, the sequences $\mathrm{Rep}(G/H) \hookrightarrow \mathrm{Rep}(G) \twoheadrightarrow \mathrm{Rep}(H)$ and $H \hookrightarrow G \twoheadrightarrow G/H$ are *split* exact sequences, owing to the induced representation $Ind_H^G : \mathrm{Rep}(H) \longrightarrow \mathrm{Rep}(G)$ and to the "Wigner rotation" as its dual $(Ind_H^G)^* : G \ni g \longmapsto (Ind_H^G)^*(g) \in H$, respectively:

$$\mathrm{Rep}(G/H) \hookrightarrow \mathrm{Rep}(G) \overset{Ind_H^G}{\leftrightarrows} \mathrm{Rep}(H),$$

$$H \underset{(Ind_H^G)^*}{\rightleftarrows} G \twoheadrightarrow G/H.$$

Namely, the vector bundle Rep$(G)$ on the base space Rep$(H)$ with standard fiber Rep$(G/H)$ has a horizontal lift given by $Ind_H^G$ and the principal $H$-bundle $G$ over $G/H$ has a $H$-valued connection given by $(Ind_H^G)^*$. In this way, the induced representation can be viewed as a theoretical basis of the structural ananlogy with the gauge theory in terms of the gauge connection $(Ind_H^G)^*$ as a splitting of exact sequence.

In the case of (2) with the Coulomb mode, we need not worry about the appearance of indefinite metric because the longitudinal Coulomb mode of classical gauge fields is described in terms of the commutative variables. Instead, what can be non-trivial now is the possibility for condensed modes of particles due to the Coulomb attractive force, according to which such non-trivial effects as superconductivity phenomena can be realized.

## 4.12 Spacelike Momenta Shared by Statistical Mechanics, Regge Poles, Dressed Photons and Coulomb Force

After the case studies of (1) induced representations and (2) gauge theories with Coulomb mode, what to be analyzed for the purpose of understanding the common features among various composite systems with inclusion relations can be found as follows:

(3) statistical mechanics and thermodynamics:

(4) Regge trajectories appearing in hadron scattering processes,

(5) mechanism of dressed photons.

Because of the big difference in the appearance among these five cases, however, it may be unclear where we can find any coherent common features. Just skipping the detailed account along individual specific features, the common essence shared by all these cases can be found in the existence of the following three levels as well as their mutual relationship:

(a) a compact Lie group $H$ to describe the inivisible microscopic dynamics associated with some flows,

(b) the level of "*horizontal duality*" formed by the algebra $\mathcal{X}$ of observables to visualize $H$ and the state space $E_{\mathcal{X}}(\subset \mathcal{X}^*)$ of $\mathcal{X}$ which is controlled by a Lie group $G$ containing $H$ as a subgroup, and,

(c) the sector classifying space $Spec(\supset G/H)$ emerging from the states $E_{\mathcal{X}}$ of $\mathcal{X}$.

What is most important is such a situation that the group $G(\supset H)$ controlling the level (b) of "*horizontal duality*" is a non-compact Lie group with a Killing form with indefinite signature, arising from the extention of the group $H$ of microscopic dynamics, characterized by the exact sequences:

$$H \hookrightarrow G \twoheadrightarrow G/H,$$

$$\mathrm{Rep}(G/H) \hookrightarrow \mathrm{Rep}(G) \twoheadrightarrow \mathrm{Rep}(H).$$

For instance, in the case of dressed photons, the region with ***spacelike momenta*** is created by introducing the ***Clebsch-dual variables*** and in the case of ***Regge trajectories*** in hadron physics, the $t$ and $u$- ***channels*** formed via the duality transformations $s \rightleftarrows t$ and $s \rightleftarrows u$ interchanging $s$, $t$ and $u$-***channels*** provide the stages of ***Regge trajectories*** consisting of the series of ***Regge poles with complex angular momenta.*** While the well-known Gibbs formula $\langle A\rangle = Tr(Ae^{-\beta H})/Tr(e^{-\beta H})$ in statistical mechanics shows no remarkable structural features, it can be applied only to small finite systems with discrete energy spectrum. In contrast, ***Tomita-Takesaki modular theory*** required for the treatment of general systems with infinite degrees of freedom is equipped with such a double structure as consisting of the von Neumann algebra $\mathcal{M}$ of physical variables in the system and its modular dual $\mathcal{M}' = J\mathcal{M}J$ whose composite system $\mathcal{M}\vee J\mathcal{M}J$ is controlled by the Hamiltonian $H_\beta = -JH_\beta J$ with "indefinite metric", whose physical interpretation can be reduced to the concept of heat bath.

## 4.13 Induced Representations and Automorphic Forms

The induced representation $Ind_H^G(\sigma)$ of the Lorentz group $G = SL(2, \mathbb{C})$ determined by a unitary representation $\sigma$ of the rotation group $H = SU(2)$ in a finite-dimensional vector space $W$ is given in an infinite-dimensional Hilbert space $V$ defined by

$$V := \{\varphi : G \longrightarrow W; \varphi(gh) = \sigma(h^{-1})\varphi(g) \text{ for } g \in G, h \in H\}$$

according to the defining equation $[Ind_H^G(\sigma)(g)\varphi](g_1) := \varphi(g^{-1}g_1)$, which reproduces $\sigma(h)$ for $h \in H$ at $g = e \in G$:

$$[Ind_H^G(\sigma)(h)\varphi](e) = \sigma(h)[\varphi(e)].$$

By means of the horizontal lift $G/H \longrightarrow G$ of $G/H = SL(2, \mathbb{C})/SU(2)$ associated with the "Wigner rotation" $(Ind_H^G)^*$, the domain of $Ind_H^G(\sigma)$ can be shifted from $G$ to $G/H$. Therefore, if we express the elements $g \in G$ in the form of fractional linear transformation, the above definition of $V$ can be rewritten as

$$V = \{\varphi : G/H \to W; \varphi(gz) = \sigma(cz+d)^{-1}\varphi(\frac{az+b}{cz+d}),$$

$$g = \begin{pmatrix} a & b \\ c & d \end{pmatrix} \in G, z \in G/H\},$$

which shows that the module $V$ consists of automorphic forms $\varphi$. Since automorphic forms are transformed into $\zeta$ functions by Mellin transform, the pair $(G, H)$ with $G/H$ a symmetric space is related to the number-theoretical contexts. While the use of fractional linear transformation: $gz := \frac{az+b}{cz+d}$ for $g = \begin{pmatrix} a & b \\ c & d \end{pmatrix} \in G$ may look accidental owing to the $(2 \times 2)$-matricial form of $SL(2, \mathbb{C})$, this is not the case because this speciality can be easily lost by such identification of the Lorentz group as $G \simeq SO(1,3) \hookrightarrow M(4, \mathbb{R})$. Actually, what is essential here is not such a special form of matrices but the orthogonal decomposition of the representation vector space $\mathfrak{V}$ of $G$ into the unbroken $\mathfrak{V}_1$ and broken subspaces $\mathfrak{V}_2$, $\mathfrak{V} = \mathfrak{V}_1 \oplus \mathfrak{V}_2$, according to which $G$ has such an orthogonal decomposition as $\begin{array}{c} \\ \mathfrak{V}_1 \\ \mathfrak{V}_2 \end{array}\!\!\begin{array}{c} \mathfrak{V}_1 \; \mathfrak{V}_2 \\ \begin{pmatrix} H & B \\ C & D \end{pmatrix} \end{array}$ in a certain neighbourhood of the identity element of $G$. Moreover, if we want to extend the above bipolar contrast between unbroken vs broken into some scale-dependent multi-polar gradations of symmetry breakings along many steps, we can consider such a flag manifold structure as related with a multi-component decomposition $\mathfrak{V} = \mathfrak{V}_1 \oplus \mathfrak{V}_2 \oplus \cdots \oplus \mathfrak{V}_r$ of the representation space $\mathfrak{V}$:

$$\begin{aligned} G &= U(p_1 + p_2 + \cdots + p_r) \\ &\curvearrowright G/H = U(p_1 + p_2 + \cdots + p_r)/[U(p_1) \times U(p_2) \times \cdots \times U(p_r)], \end{aligned}$$

which may be related with the continued fractions. In this context, we note the remarkable features in the coupling pattern of four forces as follows.

## 4.14 Upper Triangular Structure in Coupling Pattern of Four Forces

In relation with the above multi-component structures, we can see that the unification of four forces just corresponds to the evolution of dynamical laws in "historical evolution of physical nature", in the following sense.

We start from the ***duality*** of [groupoid $\Gamma$ as dynamics to deform everything] ***versus*** [***unit space*** $\Gamma^{(0)}$ as ***stabilized domain*** in Spec], in the latter of which ***natural constants*** are contained as ***deformation parameters***:

| | | |
|---|---|---|
| | Gravitational force $(\kappa, m)$ | |
| Electromagnetic force $(1/c, e)$ | $\overset{k_B}{\underset{\hbar}{\leftrightarrows}}$ | Weak forces $(e_W, \theta_{WS})$ |
| | Strong forces $(e_S, \theta_{CKM})$ | |

Here we have objects: $\varphi(\mathrm{i}) = \mathcal{M}_i$ for the indices i = g, e, h according to gravitational, electromagnetic and hadronic strong forces. In the case of i = w, we should put $\varphi(\mathrm{w}) = \mathcal{M}_\nu \otimes \mathcal{M}_W \otimes \mathcal{M}_l$ in view of the universal actions of weak forces. For morphisms we take:

$$\begin{cases} \varphi(\mathrm{g}) \curvearrowright \mathcal{M}_g \otimes \mathcal{M}_e \otimes [\mathcal{M}_\nu \otimes \mathcal{M}_W \otimes \mathcal{M}_l] \otimes \mathcal{M}_h \\ \varphi(\mathrm{w}) \curvearrowright \qquad\qquad [\mathcal{M}_\nu \otimes \mathcal{M}_W \otimes \mathcal{M}_l] \otimes \mathcal{M}_h \\ \varphi(\mathrm{e}) \curvearrowright \qquad\qquad\qquad [\mathcal{M}_W \otimes \mathcal{M}_l] \otimes \mathcal{M}_h \\ \varphi(\mathrm{s}) \curvearrowright \qquad\qquad\qquad\qquad \mathcal{M}_h \\ \varphi(\mathrm{i} \leftarrow \mathrm{j}) = [\varphi(\mathrm{i}) \leftarrow \varphi(\mathrm{j})]. \end{cases}$$

The composition series of dynamics and matter can be seen as follows:

| Dyn \ Alg | gravit'nal field | el.mg. field | neutr-inos $\nu$ | charged leptons | hadrons |
|---|---|---|---|---|---|
| g(ravity) force | ♦ | ♦ | * | * | * |
| weak force | 0 | ? | ♦ | * | * |
| el.mg. force | 0 | 0 | 0 | ♦ | * |
| strong force | 0 | 0 | 0 | 0 | * |

*upper* : *triangular matrix!!*

In the direction $\nwarrow$ from the right bottom to the left top, we have the ***historical formation of hierarchical domains*** with the first appearance ♦ of ***stabilized forms of matter*** and with the associated establishment of ***repeatable laws***.

In the opposite direction $\searrow$ from the left top to the right bottom, we can find the direction to deepen the human ***recognitions***.

The divergent arrows $\nwarrow\uparrow$ and the convergent ones $\overleftarrow{\nwarrow}$ constitute historical path of nature.

The above ? shows different behaviours of electromagnetic field at different scales:

$$? = \begin{bmatrix} 0 : \text{behaviours as purely abelian gauge field in } \textbf{\textit{Macro}} \\ \blacklozenge : \textbf{\textit{Weinberg-Salam mixing}} \text{ of } (e) \leftrightarrows (w) \text{ in } \textbf{\textit{Micro}} \end{bmatrix}.$$

The bottom horizontal row $\big(s(\mathrm{trong})|0\ 0\ 0\ 0\ *\big)$ means the absence of strong force carriers other than ***highly active hadrons*** with ***persistent existence*** under all forces in $\begin{pmatrix} \underline{\text{hadrons}} \\ * \\ * \\ * \\ * \end{pmatrix}$, which is to be contrasted with the first horizontal row

$\left(\text{gravity}\,\middle|\, * * * * *\right)$ meaning the ***universality*** of the gravitational force.

The above ***upper triangular*** structure: $\begin{array}{|c|c|}\hline \mathcal{A} \nwarrow & F \\ \hline 0 & \mathcal{X} \\ \hline \end{array}$, can be understood mathematically in terms of a ***flag manifold*** (with ***Borel subalgebra***) describing a nested solvability structure of ***composition series***.

We can interpret the upper off-diagonal term $F$ as a composite system of diagonals $\mathcal{A}$ and $\mathcal{X}$ characterized by the ***exact sequence*** $\mathcal{A} \hookrightarrow F \twoheadrightarrow \mathcal{X}$, equipped with a structure of left-$\mathcal{A}$ right-$\mathcal{X}$ ***Hilbert module*** ${}_{\mathcal{A}}F_{\mathcal{X}}$ (with $\mathcal{A} = FF^*$-valued left inner product and $\mathcal{X} = F^*F$-valued right inner product). $F$ can also be regarded as a ***left-adjoint functor*** $\mathcal{A} \overset{F}{\longleftarrow} \mathcal{X}$.

Thus, our upper triangularity with 0 in the lower left corner implies the ***presence of uni-directional arrows*** $\nwarrow$, which provide a monotone scaling parameter (similarly to Lieb-Yngvason's axiomatic derivation of entropy from adiabatic ordering) of what can be interpreted as ***historical uni-directionality inherent already at Micro-level***.

***(1) spacetime emergence: Einstein equation*** with $T_{\mu\nu}$ and gravitational force $\Gamma^{\lambda}_{\mu\nu}$ among $(g)$, $(e)$, $(w)$, $(s)$.

***(2) Maxwell equation*** and electromagnetic force $(e)$ from $(w)$, $(s)$.

Dual of the above ***co-emergence***: (***Alg*** $\curvearrowleft$ ***Dyn***), is the ***emergence***: $(\textbf{\textit{Alg}} \curvearrowleft \textbf{\textit{Dyn}})^* =$ (***Spec*** $\Leftarrow$ ***States***):

| Spec \ States | | gravit'.al field confg | el.mg. field | $\nu$ | charged leptons | hadronic states |
|---|---|---|---|---|---|---|
| gravit'.al force $\Gamma^{\lambda}_{\mu\nu}$ $\curvearrowright$ $\{x^{\mu}\}$ = index of emergence | : | terminal objects ♦ | * | * | * | * |
| weak force $\Leftarrow I_3$ | | 0 | ?* | ♦ | * | * |
| el.mg.force $\Leftarrow Q$ | | 0 | 0 | 0 | ♦ | * |
| strong force $\Leftarrow (I, B, S)$ of Regge traj.'s | | 0 | 0 | 0 | 0 | *: initial obj.'s |

## 4.15 "Indefinite Metric" Inherent in Modular Structure of Thermal Equilibrium

The above section title represents an observation to reveal a blind spot in the "common sense" in physics which can interpret the "stability" of a state only in such a restricted form as the poisitivity of the energy in the form of spectral condition.

While, in the infinite system with the operator $e^{-\beta H}$ out of trace class, it is impossible to separate sharply the physical system and its heat bath, the mutual relation between them can be mathematically understood [5, 23] by the relation:

$$H_\beta = -JH_\beta J. \tag{4.1}$$

If the component $H$ of $H_\beta$ acting on the system $\mathcal{X}_\omega$ can safely be extracted and be separated from that on the commutant $\mathcal{X}'_\omega$, then the essential contents of this equation could be seen in such a form as

$$H_\beta = H - JHJ,$$

In infinite systems, however, the above $H$ can be meaningful only formally. Apart from this subtlety, the above formal equation explains such essence that the anti-unitary operator $J$ takes care of the interchange between the system and its heat bath. Since the total system consisting of the system and the heat bath has the Hamiltonian $H_\beta$ whose spectrum is positive/negative symmetric as shown in (4.1), the negative energy component may be interpreted as the energy going from the system to the heat bath in the energy exchange processes between the system and the heat bath. Interestingly enough, the concept of "heat bath" which is mysterious but important in thermodynamics has once been driven away by the Gibbs formula (which is applicable only for the system with discrete spectrum), but, has survived in the abstract form in the algebraic general formulation of statistical mechanics based upon the so-called K(ubo-)M(artin-)S(chwinger) condition [5, 23, 24]:

$$\omega_\beta(AB(t)) = \omega_\beta(B(t - i\beta)A),$$

which is free from such a restriction of discrete energy spectrum.

Similarly to the longitudinal photons with "***negative metric***", the Hamiltonian $H_\beta$ of the total system contains the negative component (formally, $-JHJ$), which means the existence of the saddle point structure associated with the thermal equilibrium states. Without the presence of unstable modes and their condensations, the existence of the macroscopic heat bath may have been impossible.

## 4.16 Induction and Imprimitity

The following discussions are concerning how to attain the universal controls over inclusion relations in Lorentz geometry/ Regge trajectories/ Dressed photons/ Coulomb modes via induction/ restriction methods based on the concepts of induced representations and on the Tomita–Takesaki modular theory.

For the Lorentz geometry its characteristic structure will be focused from such a viewpoint that the whole group structure of the Lorentz group $G = SO(1, 3)$ (or its universal covering group $SL(2, \mathbb{C})$) can be seen as the extension of its maximal compact subgroup $H = SO(3)$ (or its universal covering group $SU(2)$) of spatial rotations via the application of Lorentz boosts forming a symmetric space $G/H$, which can be summarized in an ***exact sequence*** $H \hookrightarrow G \twoheadrightarrow G/H$ of groups and/or homogeneous spaces:

$$H \hookrightarrow G \twoheadrightarrow G/H.$$

While this special case of an exact sequence with $G = SO(1,3)$ (or $SL(2,\mathbb{C})$) and $H = SO(3)$ (or $SU(2)$)

$$SO(3) \hookrightarrow SO(1,3) \twoheadrightarrow SO(1,3)/SO(3),$$

$$SU(2) \hookrightarrow SL(2,\mathbb{C}) \twoheadrightarrow SL(2,\mathbb{C})/SU(2),$$

play crucially important roles in the following discussions, the exact sequence $H \hookrightarrow G \twoheadrightarrow G/H$ itself with *unspecified* G and $H$ will also be very important in our discussion concerning symmetry breakings, especially in close relation with such situations that a given representation $\sigma \in Rep(H)$ of $H$ is extended to that of $G$ via the method of *induced representations*: $Rep(H) \ni \sigma \longmapsto \gamma = Ind_H^G(\sigma) \in Rep(G)$ and also in the opposite direction to identify a given $G$-representation $\gamma \in Rep(G)$ as a representation $\gamma = Ind_H^G(\sigma)$ induced from an $H$-representation $\sigma \in Rep(H)$ : $Rep(G) \ni \gamma = Ind_H^G(\sigma) \longmapsto \sigma \in Rep(H)$. This latter process is taken care of by the ***imprimitivity theorem*** based on Mackey's concept of "***system of imprimitivity***" [7]. The mutual relation between these two opposite processes is controlled by the Frobenius reciprocity:

$$Rep_H(\gamma \upharpoonright_H \longrightarrow \sigma) \leftrightarrows Rep_G(\gamma \longrightarrow Ind_H^G(\sigma))$$

or

$$Rep_G(Ind_H^G(\sigma) \longrightarrow \gamma) \leftrightarrows Rep_H(\sigma \longrightarrow \gamma \upharpoonright_H),$$

where the notation $Rep_G(\gamma_1 \longrightarrow \gamma_2)$ means the set of intertwiners $T : \gamma_1 \longrightarrow \gamma_2$ from $\gamma_1$ to $\gamma_2$ as such linear operators $T$ from the representation Hilbert space $V_{\gamma_1}$ of $\gamma_1$ to that $V_{\gamma_2}$ of $\gamma_2$ as satisfying the intertwining relation $\forall g \in G\ T\gamma_1(g) = \gamma_2(g)T$, namely,

$$T \in Rep_G(\gamma_1 \longrightarrow \gamma_2) \Longleftrightarrow \forall g \in G : T\gamma_1(g) = \gamma_2(g)T$$

$$\begin{array}{ccc} V_{\gamma_1} & \xrightarrow{T} & V_{\gamma_2} \\ \gamma_1(g)\downarrow & \circlearrowleft & \downarrow \gamma_2(g) \\ V_{\gamma_1} & \xrightarrow[T]{} & V_{\gamma_2} \end{array}$$

In view of the definition for a group representation $\gamma \in Rep_G$ given by the group homomorphism properties $\gamma(g_1 g_2) = \gamma(g_1)\gamma(g_2)$, $\gamma(e) = id_{V_\gamma}$, $\gamma(g^{-1}) = \gamma(g)^{-1}$, a

$G$-representation $\gamma$ can be viewed as a ***functor*** from the group $G$ as a one-object category $G = \mathcal{C}_G$ consisting of an object $*$ and of group elements $g \in G$ as morphisms $(* \xrightarrow{g} *) \in G = Mor(\mathcal{C}_G))$ to another category $Hom(V_\gamma)$ consisting of continuous linear operators in the Hilbert space $V_\gamma$. From this categorical viewpoint, the intertwiner $T \in Rep_G(\gamma_1 \longrightarrow \gamma_2)$ from $\gamma_1$ to $\gamma_2$ is to be interpreted as a ***natural transformation*** from a functor $\gamma_1$ to another functor $\gamma_2$ characterized by the above commutativity diagram. In this way, the totality $Rep_G$ of $G$-representations can be viewed as a category $Hilb^G$ of functors from the group $G$ as a category $\mathcal{C}_G = G$ to the category $Hilb$ of Hilbert spaces with morphisms given by $G$-intertwiners as natural transformations. In this context, the group induction $Ind_H^G$ from the functor category $Rep_H$ of $H$-representations to that $Rep_G$ of $G$-representations can be viewed as a natural transformation $Ind_H^G : Rep_H \longrightarrow Rep_G$ (preserving the tensor product structures of $Rep_H$ and $Rep_G$: $Ind_H^G(\sigma_1 \otimes \sigma_2) = Ind_H^G(\sigma_1) \otimes Ind_H^G(\sigma_2)$ for $\sigma_1, \sigma_2 \in Rep_H$).

Let $\mathcal{X}$ be the algebra of physical variables of a physical system to be described. Under an action $\alpha = (\alpha_g)_{g \in G}$ of a Lie group $G$ via automorphisms $\alpha_g$ on the algebra $\mathcal{X}$, the observable algebra $\mathcal{A}$ is defined as the $G$-invariant subalgebra of $\mathcal{X}$ by

$$\mathcal{A} = \mathcal{X}^G := \{A; \alpha_g(A) = A \text{ for } \forall g \in G\}.$$

Under a suitable assumption, an exact sequence

$$\mathcal{A} \hookrightarrow \mathcal{X} \twoheadrightarrow \mathcal{X}/\mathcal{A} \cong \widehat{G}$$

arises in this situation, from which the total algebra $\mathcal{X}$ can be recovered from the observable algebra $\mathcal{A}$ [17, 25] by means of the crossed product of $\widehat{G}$ in the context of the categorical adjunction:

$$\mathcal{A} = \mathcal{X}^G \leftrightarrows \mathcal{X} = \mathcal{A} \lhd \widehat{G}.$$

When we combine the inclusion relation of groups controlled by the exact sequence $H \hookrightarrow G \twoheadrightarrow G/H$ with the group actions on the algebras of physical variables, we encounter the situation of symmetry breakings which involves the mutual relations among various subalgebras $\mathcal{X}^G \hookrightarrow \mathcal{X}^H \hookrightarrow \mathcal{X}$.

To attain a unified control over various mutual relations among the above sort of objects, we consider the formulation of the processes of extension and restriction in terms of the categorical functors, since groups and algebras can be regarded as special cases of categories and homomorphisms among them as functors among such categories.

## 4.17 Kan Extensions as Categorical Inductions

Given a funcor $K : \mathcal{B} \to \mathcal{A}$ from a category $\mathcal{B}$ to $\mathcal{A}$ we consider the problem of extending a given functor $S : \mathcal{B} \to \mathcal{M}$ from $\mathcal{B}$ to $\mathcal{M}$ into one $T : \mathcal{A} \to \mathcal{M}$ from $\mathcal{A}$ to $\mathcal{M}$ so as to satisfy the relation $T \circ K = S$:

$$\begin{array}{ccc} & T? & \\ \mathcal{A} & \dashrightarrow & \mathcal{M} \\ K \uparrow & \circlearrowleft & \nearrow S \\ \mathcal{B} & & \end{array}.$$

In this situation, the functor $T$ is called a ***Kan extension*** [26] of functor $S$ along functor $K$. For instance, if we identify $K : \mathcal{B} \to \mathcal{A}$ as the inclusion $\iota : H \hookrightarrow G$ of a subgroup $H$ into the total group $G$ and $S : \mathcal{B} \to \mathcal{M}$ as a representation $\sigma : H \to \mathcal{M} = Hilb$ of $H$ with $Hilb$ identified with the category of Hilbert spaces, then $T : \mathcal{A} \to \mathcal{M}$ corresponds to an extension of $H$-representation $\sigma$ to $G$-representation $\gamma$:

$$\begin{array}{ccc} & \gamma? & \\ G & \dashrightarrow & \mathcal{M} \\ \iota \uparrow & \circlearrowleft & \nearrow \sigma \\ H & & \end{array},$$

since the commutativity $\sigma = \gamma \circ \iota$ of the diagram means $\sigma = \gamma \upharpoonright_H$. In this sense, the Kan extension can be viewed as a categorical version of the induced representations of groups.

In view of the important roles played by natural transformations in mediating adjoint functors, we need to distinguish between the right and left Kan extensions as follows:

$$Nat_{\mathcal{M}^{\mathcal{B}}}(T \circ K \longrightarrow S) \leftrightarrows Nat_{\mathcal{M}^{\mathcal{A}}}(T \longrightarrow Ran_K(S))$$

$$Nat_{\mathcal{M}^{\mathcal{A}}}(Lan_K(S) \longrightarrow T) \leftrightarrows Nat_{\mathcal{M}^{\mathcal{B}}}(S \longrightarrow T \circ K)$$

The concept of ***Yoneda embedding***:

$$\mathbf{y}_{\mathcal{C}}(-) = \mathcal{C}((-) \longrightarrow c) \in Sets^{\mathcal{C}^{op}} : \mathcal{C} \ni d \longmapsto \mathcal{C}(d \longleftarrow c) \in Sets$$

gives an embedding of a category $\mathcal{C}$ into the category $Sets^{\mathcal{C}^{op}}$ of pre-sheaves on $\mathcal{C}$ (as a categorical generalization of the concept of functions), and hence, it would be quite useful to consider the Kan extensions $Ran_{\mathbf{y}_c}$ or $Lan_{\mathbf{y}_c}$ along $K = \mathbf{y}_{\mathcal{C}}$. However, systematic investigation on this topic should be done on the next occasions.

## 4.18 Towards Theory of Dressed Photons

In order to construct a consistent theory for describing dressed photons, it will become a crucial breakthrough to reproduce faithfully its proper dynamic functions by grasping properly the "tapering" cone structure formed by the condensed dressed photons. To implement the ideas in this direction, it is important to install the ***Clebsch-dual electromagnetic field*** [18] discovered by Sakuma carrying spacelike momenta which constitute the characteristic off-shell structure of electromagnetic field. which forms the Micro–Macro boundary level described by a symmetric space $G/H = Spec$ arising from a broken symmetry by visualizing the $s$-channel structure at the invisible Micro level into spacelike $t$-channel. If we succeed in identifying suitable automorphic forms defined on this symmetric space, it will become possible for us to describe any of non-trivial dynamic phenomena caused by the dressed photons by means of them. Especially, the "automorphic factor" $\sigma(cz + d)^{-1}$ appearing in the definition of an automorphic form:

$$\varphi(gz) = \sigma(cz + d)^{-1}\varphi(\frac{az + b}{cz + d}),$$

will play important roles as a ***cocycle*** carrying the dynamic properties of invisible Micro level. It will perhaps justify in the context of dressed photons the analogy with the dynamic behaviour played by the Regge trajectories carrying spacelike momenta in the hadronic scattering processes originating from the dual resonance structure.

**Acknowledgements** I would like to express my sincere thanks to Professor Ohtsu for inviting me into the exciting and challenging research field of dressed photons. Without this extremely nice opportunity, I could not find the interesting directions to which the framework of Micro–Macro duality and quadrality scheme should be developed further in such a fascinating forms of connections with experimental situations. On this opportunity, let me express my thanks to Dr. Sakuma for inspiring discussions on Clebsch duality for many years. On many subjects, I have benefited very much from the collaborations and instructive discussions with all the members at RODreP. Last but not least, I would like to thank Professor Yatsui for all his patient helps in facilitating my slow writing process.

## References

1. I. Ojima, A unified scheme for generalized sectors based on selection criteria Order parameters of symmetries and of thermal situations and physical meanings of classifying categorical adjunctions. Open Sys. Info. Dyn. **10**, 235–279 (2003)

2. I. Ojima, *Quantum Fields and Micro-Macro Duality* (in Japanese) (Maruzen, 2013)
3. I. Ojima, K. Okamura, *Physics and Mathematics of Infinite Quantum Systems* (in Japanese) (Science-Sha, SGC98, 2013)
4. I. Ojima, Micro-macro duality in quantum physics, 143–161, in *Proceedings of the International Conference "Stochastic Analysis: Classical and Quantum –Perspectives of White Noise Theory"* ed. by T. Hida (World Scientific, Singapore, 2005). arXiv:math-ph/0502038
5. O. Bratteli, D.W. Robinson, *Operator Algebras and Quantum Statistical Mechanics*, Vols. 1 and 2 (Springer, Berlin, 1979, 1981)
6. J. Dixmier, *C*-Algebras*, North-Holland, 1977; Pedersen, G., *C*-Algebras and Their Automorphism Groups* (Academic, Massachusetts, 1979)
7. G.W. Mackey, *Induced Representations of Groups and Quantum Mechanics* (W.A.Benjamin, Inc., 1968)
8. I. Ojima, Unpublished (2004)
9. Y. Nakagami, M. Takesaki, *Duality for Crossed Products of von Neumann Algebras*. Lecture Notes in Mathematics, vol. 731 (Springer, Berlin, 1979)
10. S. Helgason, *Differential Geometry, Lie Groups, and Symmetric Spaces* (Academic, New York, 1978)
11. I. Ojima, Dynamical relativity in family of dynamics. RIMS Kôkyôroku **1921**, 73–83 (2014); Local gauge invariance and Maxwell equation in categorical QFT. RIMS Kôkyôroku **1961**, 81–92 (2015); Local Gauge invariance, Maxwell equation and symmetry, talk at NWW2015; Algebraic QFT and local Gauge invariance. RIMS Kôkyôroku **2010**, 78–88 (2016)
12. R. Haag, *Local Quantum Physics*, 2nd. edn. (Springer, Berlin, 1996)
13. I. Ojima, Lorentz invariance versus temperature in QFT. Lett. Math. Phys. **11**, 73–80 (1986)
14. I. Ojima, Space(-time) emergence as symmetry breaking effect. Quantum Bio-Inf. **IV**, 279–289 (2011). arXiv:math-ph/1102.0838; Micro-Macro Duality and space-time emergence, *Proceedings of the International Conference "Advances in Quantum Theory"* (2011), pp. 197–206
15. I. Ojima, Temperature as order parameter of broken scale invariance. Publ. RIMS (Kyoto Univ.) **40**, 731–756 (2004). arXiv:math-ph/0311025
16. S. Helgason, *The Radon Transform* (Birkhäuser, 1980)
17. S. Doplicher, J.E. Roberts, Why there is a field algebra with a compact gauge group describing the superselection structure in particle physics. Commun. Math. Phys. **131**, 51–107 (1990)
18. H. Sakuma, I. Ojima, M. Ohtsu, Dressed photons from the viewpoint of photon localization: the entrance to the off-shell science. Appl. Phys. **A123**, 724 (2017); Gauge symmetry breaking and emergence of Clebsch-dual electromagnetic field as a model of dressed photons. Appl. Phys. **A123**(750) (2017). https://doi.org/10.1007/s00339-017-1364-9; Dressed photons in a new pradigm of off-shell quantum fields. Prog. Quantim Electron. **55**, 74–87
19. S. Doplicher, R. Haag, J.E. Roberts, Fields, observables and gauge transformations I and II. Commun. Math. Phys. **13**, 1–23 (1969); **15**, 173–200 (1969); Local observables and particle statistics, I and II, **23**, 199–230 (1971); **35**, 49–85 (1974)
20. T. Kugo, I. Ojima, Local covariant operator formalism of non-Abelian Gauge theories and quark confinement problem, Suppl. Prog. Theor. Phys. No. **66** (1979); N. Nakanishi, I. Ojima, *Covariant Operator Formalism of Gauge Theories and Quantum Gravity*. Lecture Notes in Physics, vol. 27 (World Scientific Publishing Company, Singapore, 1990)
21. J. Milnor, *Morse Theory* (Princeton University Press, Princeton, 1963)
22. E.P. Wigner, On unitary representations of the inhomogeneous Lorentz group. Ann. Math. **40**, 149–204 (1939)
23. R. Haag, N.M. Hugenholtz, M. Winnink, On the equilibrium states in quantum statistical mechanics. Commun. Math. Phys. **5**, 215–236 (1967)
24. R. Kubo, J. Phys. Soc. Japan **12**, 570–586 (1957); P.C. Martin, J. Schwinger, Theory of many particle systems I. Phys. Rev. **115**, 1342–1373 (1959)
25. S. Doplicher, J.E. Roberts, Endomorphism of C*-algebras, cross products and duality for compact groups. Ann. Math. **130**, 75–119 (1989); A new duality theory for compact groups. Inventiones Math. **98**, 157–218 (1989)
26. See, for instance, S. MacLane, *Categories for the Working Mathematician* (Springer, Berlin, 1971)

# Chapter 5
# An Approach from Measurement Theory to Dressed Photon

**Kazuya Okamura**

**Abstract** In this review, we discuss quantum measurement theory to develop that for dressed photons. Algebraic quantum theory is based on the theory of operator algebras, especially, C*-algebras. Our formulation of quantum measurement theory stands on von Neumann algebras, a special class of C*-algebra, and treats processes of measurements both in the Schrödinger picture and in the Heisenberg picture.

## 5.1 Introduction

In this review, we discuss quantum measurement theory in algebraic quantum theory to develop that for dressed photons. Algebraic quantum theory is based on the theory of operator algebras, especially, C*-algebras. The advantage of C*-algebraic quantum theory is that it can explicitly describe macroscopic classical levels of quantum systems. A sector, a quasi-equivalence class of a factor state, is a macroscopic unit of quantum system. And we can describe the statistical aspect of sectors in the measure-theoretical probability theory. In terms of sectors, we axiomatically formulate algebraic quantum theory and discuss symmetry breaking.

Our formulation of quantum measurement theory stands on von Neumann algebras, a special class of C*-algebra, and is applicable to local measurements in algebraic quantum field theory. In the Schrödinger picture, processes of measurements are described by the notion of completely positive instrument defined on von Neumann algebras. On the other hand, we can consider quantum mechanical modelings of measuring appratuses, called measuring processes. It is shown that there exists a one-to-one correspondence between completely positive instruments with "the normal extension property" and "statistical equivalence classes" of measuring

K. Okamura (✉)
Graduate School of Infomatics, Nagoya University, Furo-cho, Chikusa-ku, Nagoya, Aichi 464-8601, Japan
e-mail: okamura@math.cm.is.nagoya-u.ac.jp



T. Yatsui (ed.), *Progress in Nanophotonics 5*, Nano-Optics and Nanophotonics,
https://doi.org/10.1007/978-3-319-98267-0_5

processes. Therefore, two approach to measurement theory is almost equivalent. This equivalence is justified by the result that every completely instrument can be approximated by completely positive instruments with the normal extension property. In contrast to the Schrödinger picture, processes of measurements in the Heisenberg picture are described by the notion of system of measurement correlations. Systems of measurement correlations are characterized like Wightman's axiomatic system of quantum fields [1]. Every completely positive instrument can be extended into a system of measurement correlations.

To formulate the (universally valid) theory of dressed photons [2], we have several difficulties to overcome. One of them is measurement theory for dressed photons. It is usually said that processes of measurements affect the system to be measured. For dressed photons, we can make a stronger statement; They are exactly the system which can never be measurable without disturbing their dynamics as long as they are in the measurement situation. We cannot present measurement theory for dressed photons herein. However, I wrote this manuscript for the theoretical preparation.

## 5.2 Algebraic Quantum Theory and Sector Theory

To develop the discussion of the review, it is essential to construct the theoretical basis of algebraic quantum theory. Algebraic quantum theory based on operator algebras can use mathematical concepts of many areas and can explicitly describe macroscopic classical levels of quantum systems usually excluded from quantum mechanics. Generalized sectors defined by Ojima [3–5] play the role of macroscopic unit. Classifying spaces as spectra (Spec) are constructed from Families of sectors. By contrast, both symmetry which controls mutual relations among sectors, and time evolution are the essentials of the dynamics (Dyn) of the system. For the description of Spec and Dyn, we need algebras (Alg) of observables and states (States) on algebras corresponding to experimental settings. The quadruplet of Alg, States, Spec and Dyn form a quadrality scheme advocated by Ojima [6]. We will see the potential of algebraic quantum theory using sectors.

In this section, foundations of operator algebras, an axiomatic system of algebraic quantum theory, and symmetry breaking are discussed. We examine von Neumann's axiomatic system of quantum mechanics thereafter.

### *5.2.1 Preliminaries on Operator Algebras*

Here, we introduce basic facts on operator algebras. We refer the reader to [7–15] for details on operator algebras and their applications.

**Definition 1** (*$C^*$-algebra*) A (complex) Banach space $\mathcal{X}$ is called a $C^*$-algebra if it satisfies the following conditions:

(1) $\mathcal{X}$ is $^*$-algebra, i.e., $\mathcal{X}$ is an algebra which has the involution $^* : \mathcal{X} \to \mathcal{X}$ satisfying $(XY)^* = Y^*X^*, (X+Y)^* = X^* + Y^*, X^{**} := (X^*)^* = X$ and $(aX)^* = \bar{a}X^*$ for all $X, Y \in \mathcal{X}$ and $a \in \mathbb{C}$.
(2) $\|X^*X\| = \|X\|^2$ for all $X \in \mathcal{X}$.

An element $X$ of $\mathcal{X}$ is said to be self-adjoint if $X^* = X$. Two elements $X$ and $Y$ are (mutually) commutative if $XY = YX$. A C$^*$-algebra $\mathcal{A}$ is said to be commutative or abelian if $XY = YX$ for all $X, Y \in \mathcal{X}$. For every Hilbert space $\mathcal{H}$, the set $\mathbb{B}(\mathcal{H})$ of bounded linear operators on $\mathcal{H}$ is a C$^*$-algebra.

**Definition 2** (*State*) A (complex) linear functional $\omega$ on $\mathcal{X}$ is called a state on $\mathcal{X}$ if it satisfies the following conditions:
(1) $\omega(X^*X) \geq 0$ for all $X \in \mathcal{X}$.
(2) $\|\omega\| := \sup_{\|X\| \leq 1,\, X \in \mathcal{X}} |\omega(X)| = 1$.
If $\mathcal{X}$ is unital, then the condition (2) is equivalent to $\omega(1) = 1$.

For every C$^*$-algebra $\mathcal{X}$, $\mathcal{S}(\mathcal{X})$ denotes the set of states on $\mathcal{X}$. A pair $(\mathcal{X}, \omega)$ of a C$^*$-algebra $\mathcal{X}$ and a state $\omega$ on $\mathcal{X}$ is called a C$^*$-probability space.

Let $(S, \mathcal{F}, \mu)$ be a measure space. $\mathcal{L}(S, \mu)$ denotes the $^*$-algebra of complex-valued $\mu$-measurable functions on $S$ and $M^\infty(S, \mu)$ does the $^*$-algebra of bounded complex-valued $\mu$-measurable functions on $S$. A $\mu$-measurable function $f$ is said to be $\mu$-negligible if $f(s) = 0$ for $\mu$-a.e. $s \in S$. $\mathcal{N}(S, \mu)$ denotes the set of $\mu$-negligible functions on $S$. For any $1 \leq p < \infty$, $L^p(S, \mu)$ denotes the Banach space of $p$-integrable functions on $S$ with respect to $\mu$ modulo the $\mu$-negligible functions. We define the Banach space $L^\infty(S, \mu)$ by $L^\infty(S, \mu) = M^\infty(S, \mu)/\mathcal{N}(S, \mu)$. $[f]$ denotes the $\mu$-negligible equivalence class of $f \in \mathcal{L}(S, \mu)$ and $L^\infty(S, \mu)$ does the commutative von Neumann algebra on $L^2(S, \mu)$.

**Definition 3** (*von Neumann algebra and W$^*$-algebra*) (1) Let $\mathcal{H}$ be a Hilbert space. A $^*$-subalgebra $\mathcal{M}$ of $\mathbb{B}(\mathcal{H})$ is called a von Neumann algebra (on $\mathcal{H}$) if it is (ultra-) weakly closed in $\mathbb{B}(\mathcal{H})$.

(2) A C$^*$-algebra $\mathcal{N}$ is called a $W^*$-algebra if there exists a Banach space $\mathcal{G}$ such that $\mathcal{N}$ is isomorphic to $\mathcal{G}^*$ as Banach space.

The Banach space $G$ is unique by Sakai's theorem and is called the predual of $\mathcal{M}$, denoted by $\mathcal{M}_*$. It is known that $\mathcal{M}_*$ coincides with the set of ultraweakly continuous linear functionals on $\mathcal{M}$. $\mathbb{B}(\mathcal{H})$ is a von Neumann algebra and there exists a one-to-one correspondence between elements of $\mathbb{B}(\mathcal{H})_*$ and elements of the set $\mathbb{T}(\mathcal{H})$ of trace class operators on $\mathcal{H}$, where the one-to-one correspondence $\mathbb{B}(\mathcal{H})_* \ni \rho \leftrightarrow \sigma \in \mathbb{T}(\mathcal{H})$ is given by

$$\langle X, \rho \rangle = \mathrm{Tr}[\sigma X] \tag{5.1}$$

for all $X \in \mathbb{B}(\mathcal{H})$.

**Definition 4** (*$^*$-isomorphism and $W^*$-isomorphism*) (1) Let $\mathcal{X}, \mathcal{Y}$ be C$^*$-algebras. A homomorphism $j : \mathcal{X} \to \mathcal{Y}$ is called a $^*$-homomorphim if $j(X^*) = j(X)^*$ for

all $X \in \mathcal{X}$. A $^*$-homomorphim $j : \mathcal{X} \to \mathcal{Y}$ is called a $^*$-isomorphism if it is one-to-one. We say that $\mathcal{X}$ is $^*$-isomorphic to $\mathcal{Y}$, denoted by $\mathcal{X} \cong \mathcal{Y}$, if there exists a $^*$-isomorphism $j : \mathcal{X} \to \mathcal{Y}$.
(2) Let $\mathcal{M}$, $\mathcal{N}$ be $W^*$-algebras. A $^*$-homomorphism $j : \mathcal{M} \to \mathcal{N}$ is called a $W^*$-homomorphim if it is normal, i.e., for every upper bounded increasing net $\{M_\alpha\}_\alpha$ of positive elements of $\mathcal{M}$ in the $\sigma(\mathcal{M}, \mathcal{M}_*)$ topology, $\{j(M_\alpha)\}_\alpha$ is an upper bounded increasing net of $\mathcal{N}$ in the $\sigma(\mathcal{N}, \mathcal{N}_*)$ topology. A $W^*$-homomorphim $j : \mathcal{X} \to \mathcal{Y}$ is called a $W^*$-isomorphism if it is one-to-one. We say that $\mathcal{M}$ is $W^*$-isomorphic to $\mathcal{N}$, denoted by $\mathcal{M} \cong \mathcal{N}$, if there exists a $W^*$-isomorphism $j : \mathcal{M} \to \mathcal{N}$.

**Definition 5** (*$^*$-representation*) Let $\mathcal{X}$ be a C$^*$-algebra and $\mathcal{H}$ a Hilbert space. We call a $^*$-homomorphism $\pi : \mathcal{X} \to \mathbb{B}(\mathcal{H})$ a $^*$-representation of $\mathcal{X}$ on $\mathcal{H}$ (or simply a representation of $\mathcal{X}$). $(\pi, \mathcal{H})$ denotes a representation $\pi$ of a C$^*$-algebra $\mathcal{X}$ on a Hilbert space $\mathcal{H}$.

Any abelian C$^*$-algebra admits the following representation theorem.

**Theorem 1** (Gel'fand–Naimark) *Let $\mathcal{A}$ be an abelian C$^*$-algebra. There exists a locally compact Hausforff space $S$ such that $\mathcal{A} \cong C_0(S)$.*

**Theorem 2** ([14, Chap. III, Theorem 1.18]) *Let $\mathcal{M}$ be an abelian von Neumann algebra on a Hilbert space. There exist a locally compact Hausforff space $S$ and a regular Borel measure on $S$ such that $\mathcal{M}$ is $W^*$-isomorphic to $L^\infty(S, \mu)$.*

The following theorem states that every positive linear functional on $C_0(S)$ is nothing but a regular Borel measure on $S$.

**Theorem 3** (Riesz–Markov–Kakutani) *Let $S$ be a locally compact Hausforff space. There exists a one-to-one correspondence between a positive linear functional $\rho$ on $C_0(S)$ and a regular Borel measure $\mu$ on $(S, \mathcal{B}(S))$, which is given by*

$$\rho(f) = \int_S f(s)\, d\mu(s) \tag{5.2}$$

*for all $f \in C_0(S)$.*

Furthermore, the following theorem states that every abstract C$^*$-algebra is $^*$-isomorphic to a concrete C$^*$-algebra.

**Theorem 4** (Gel'fand–Naimark) *For every C$^*$-algebra $\mathcal{X}$, there exist a Hilbert space $\mathcal{H}$ and a C$^*$-subalgebra $\mathcal{Y}$ of $\mathbb{B}(\mathcal{H})$ such that $\mathcal{X} \cong \mathcal{Y}$.*

Next, we shall consider representations of C$^*$-algebras for each state.

**Definition 6** (*GNS representation*) Let $\mathcal{X}$ be a C$^*$-algebra. A triplet $(\pi, \mathcal{H}, \Omega)$ is called a GNS representation of a positive linear functional $\omega$ on $\mathcal{X}$ if $(\pi, \mathcal{H})$ is a representation of $\mathcal{X}$ and $\Omega \in \mathcal{H}$ and it holds that

$$\omega(X) = \langle \Omega | \pi(X)\Omega \rangle \tag{5.3}$$

for all $X \in \mathcal{X}$ and $\mathcal{H} = \overline{\mathrm{span}}(\pi(\mathcal{X})\Omega)$,

**Definition 7** (*Unitary equivalence*) Let $\mathcal{X}$ be a C*-algebra. Let $\pi_1$ and $\pi_2$ be representations of $\mathcal{X}$ on Hilbert spaces $\mathcal{H}_1$ and $\mathcal{H}_2$, respectively. We say that $\pi_1$ is unitarily equivalent to $\pi_2$, denoted by $\pi_1 \simeq \pi_2$, if there exists a unitary operator $U \in \mathbb{B}(\mathcal{H}_1, \mathcal{H}_2)$ such that $U\pi_1(X) = \pi_2(X)U$ for all $X \in \mathcal{X}$.

We see that every state defines a unique representation of C*-algebras by the following theorem.

**Theorem 5** *For every positive linear functional $\omega$ on $\mathcal{X}$, there uniquely exists a GNS representation $(\pi, \mathcal{H}, \Omega)$ of $\omega$ up to unitary equivalence.*

Then $(\pi_\omega, \mathcal{H}_\omega, \Omega_\omega)$ denotes the GNS representation of $\omega$.

We define a Hilbert space $\mathcal{H}_u$ and the universal representation $\pi_u$ of $\mathcal{X}$ on $\mathcal{H}_u$ by

$$\mathcal{H}_u = \bigoplus_{\varphi \in \mathcal{S}(\mathcal{X})} \mathcal{H}_\varphi, \tag{5.4}$$

$$\pi_u = \bigoplus_{\varphi \in \mathcal{S}(\mathcal{X})} \pi_\varphi. \tag{5.5}$$

We define the natural embedding $j_n : \mathcal{X} \to \mathcal{X}^{**}$ by

$$\langle j_n(X), \varphi \rangle = \varphi(X) \tag{5.6}$$

for all $\varphi \in \mathcal{X}^*$.

The following two theorems show the naturality and universality of the universal representation of C*-algebras.

**Theorem 6** *Let $\mathcal{X}$ be a C*-algebra. Then the second dual $\mathcal{X}^{**}$ of $\mathcal{X}$ is a W*-algebra W*-isomorphic to $\pi_u(\mathcal{X})''$.*

**Theorem 7** *Let $\mathcal{X}$ be a C*-algebra and $\pi$ a representation of $\mathcal{X}$ on a Hilbert space $\mathcal{H}$. Then there exists a normal representation $\widetilde{\pi}$ of $\mathcal{X}^{**}$ on $\mathcal{H}$ such that $\widetilde{\pi} \circ j_n = \pi$.*

Before defining two kinds of equivalence relation of C*-algebras, we shall prepare some terminologies.

**Theorem 8** *Let $\mathcal{M}$ be a von Neumann algebra on a Hilbert space $\mathcal{H}$.*
*(1) A functional $\omega$ on $\mathcal{M}$ is ultraweakly continuous if and only if there exist sequences $\{\xi_n\}_{n\in\mathbb{N}}$, $\{\eta_n\}_{n\in\mathbb{N}}$ of $\mathcal{H}$ such that $\sum_{n=1}^{\infty} \|\xi_n\|^2 < \infty$, $\sum_{n=1}^{\infty} \|\eta_n\|^2 < \infty$ and*

$$\omega(M) = \sum_{n=1}^{\infty} \langle \xi_n | M \eta_n \rangle \tag{5.7}$$

*for all $M \in \mathcal{M}$.*
*(2) A state $\omega$ on $\mathcal{M}$ is normal if and only if there exists a sequence $\{\xi_n\}_{n\in\mathbb{N}}$ of $\mathcal{H}$ such that*

$$\omega(M) = \sum_{n=1}^{\infty} \langle \xi_n | M \xi_n \rangle \tag{5.8}$$

*for all $M \in \mathcal{M}$.*

**Definition 8** (*$\pi$-normality*) Let $\mathcal{X}$ be a C*-algebra and $\pi$ a representation of $\mathcal{X}$ on a Hilbert space $\mathcal{H}$. A state $\omega$ on $\mathcal{X}$ is $\pi$-normal if there exists a normal state $\rho$ on $\pi(\mathcal{X})''$ such that

$$\omega(X) = \rho(\pi(X)) \tag{5.9}$$

for all $X \in \mathcal{X}$. $\mathcal{S}_\pi(\mathcal{X})$ denotes the set of $\pi$-normal states.

For every $\omega \in E_{\mathcal{X}}$, $n \in \mathbb{N}$, $X_1, \ldots, X_n \in \mathcal{X}$ and $\varepsilon_1, \ldots, \varepsilon_n > 0$,

$$U_\omega(\{X_i, \varepsilon_i\}_{i=1}^n) = \{\varphi \in \mathcal{S}(\mathcal{X}) \mid |\omega(X_i) - \varphi(X_i)| < \varepsilon_i \text{ for all } i = 1, \ldots, n\}. \tag{5.10}$$

The weak*-topology of $\mathcal{S}(\mathcal{X})$ is generated by a basis $U_\omega(\{X_i, \varepsilon_i\}_{i=1}^n)$.

**Definition 9** (*Physical containment and physical equivalence*) Let $\mathcal{X}$ be a C*-algebra. Let $\pi_1$ and $\pi_2$ be representations of $\mathcal{X}$ on Hilbert spaces $\mathcal{H}_1$ and $\mathcal{H}_2$, respectively.
(1) We say that $\pi_1$ is physically contained in $\pi_2$ if $\mathcal{S}_{\pi_2}(\mathcal{X}) \cap U_\omega(\{X_i, \varepsilon_i\}_{i=1}^n) \neq \emptyset$ for every $\omega \in \mathcal{S}_{\pi_1}(\mathcal{X})$, $n \in \mathbb{N}$, $X_1, \ldots, X_n \in \mathcal{X}$ and $\varepsilon_1, \ldots, \varepsilon_n > 0$.
(2) We say that $\pi_1$ is physically equivalent to $\pi_2$ if $\pi_1$ is physically contained in $\pi_2$, and vice versa.

**Theorem 9** ([16, Chap. II, Theorem 7]) *Let $\mathcal{X}$ be a C*-algebra. Let $\pi_1$ and $\pi_2$ be representations of $\mathcal{X}$ on Hilbert spaces $\mathcal{H}_1$ and $\mathcal{H}_2$, respectively.*
*(1) $\pi_1$ is physically contained in $\pi_2$ if and only if* $\ker(\pi_1) \supset \ker(\pi_2)$.
*(2) $\pi_1$ is physically equivalent to $\pi_2$ if and only if* $\ker(\pi_1) = \ker(\pi_2)$.

**Definition 10** (*Quasi-containment and quasi-equivalence*) Let $\mathcal{X}$ be a C*-algebra. Let $\pi_1$ and $\pi_2$ be representations of $\mathcal{X}$ on Hilbert spaces $\mathcal{H}_1$ and $\mathcal{H}_2$, respectively.
(1) $\pi_1$ is said to be quasi-contained in $\pi_2$, denoted by $\pi_1 \prec \pi_2$, if there is a $W^*$-homomorphism $j : \pi_2(\mathcal{X})'' \to \pi_1(\mathcal{X})''$ such that $j \circ \pi_2 = \pi_1$.
(2) $\pi_1$ is said to be quasi-equivalent to $\pi_2$, denoted by $\pi_1 \approx \pi_2$, if there is an $W^*$-isomorphism $j : \pi_1(\mathcal{X})'' \to \pi_2(\mathcal{X})''$ such that $j \circ \pi_1 = \pi_2$.

**Theorem 10** ([14, Chap. IV, Theorem 5.5]) *Let $\mathcal{M}_1$, $\mathcal{M}_2$ be von Neumann algebras on Hilbert spaces $\mathcal{H}_1$, $\mathcal{H}_2$, respectively. Let $\pi$ be a $W^*$-homomorphism from $\mathcal{M}_1$ onto $\mathcal{M}_2$. There exist a Hilbert space $\mathcal{K}$, a projection $E$ of $\mathcal{M}' \overline{\otimes} \mathbb{B}(\mathcal{K})$ and an isometry $U$ from $E(\mathcal{H}_1 \otimes \mathcal{K})$ to $\mathcal{H}_2$ such that*

$$\pi(M) = U j_E(M \otimes 1) U^* \tag{5.11}$$

*for all $M \in \mathcal{M}_1$, where $j_E : \mathbb{B}(\mathcal{H}_1 \otimes \mathcal{K}) \to E\mathbb{B}(\mathcal{H}_1 \otimes \mathcal{K})E$ is defined by*

$$j_E(X) = EXE \tag{5.12}$$

*for all* $X \in \mathbb{B}(\mathcal{H}_1 \otimes \mathcal{K})$.

**Theorem 11** ([7, Theorem 2.4.26]) *Let* $\mathcal{X}$ *be a* $C^*$*-algebra. Let* $\pi_1$ *and* $\pi_2$ *be representations of* $\mathcal{X}$ *on Hilbert spaces* $\mathcal{H}_1$ *and* $\mathcal{H}_2$*, respectively. The following conditions are equivalent:*
(1) $\pi_1 \approx \pi_2$.
(2) $\mathcal{S}_{\pi_1}(\mathcal{X}) = \mathcal{S}_{\pi_2}(\mathcal{X})$.
(3) *There exist Hilbert spaces* $\mathcal{K}_1$, $\mathcal{K}_2$, *projections* $E_1 \in \pi_2(\mathcal{X})' \mathbin{\overline{\otimes}} \mathbb{B}(\mathcal{K}_1)$, $E_2 \in \pi_1(\mathcal{X})' \mathbin{\overline{\otimes}} \mathbb{B}(\mathcal{K}_2)$ *and isometries* $U_1 : E_1(\mathcal{H}_1 \otimes \mathcal{K}_2) \to \mathcal{H}_2$, $U_2 : E_2(\mathcal{H}_2 \otimes \mathcal{K}_1) \to \mathcal{H}_1$ *such that*

$$U_1^*\pi_1(X)U_1 = (\pi_2(X) \otimes 1)E_1, \qquad U_2^*\pi_2(X)U_2 = (\pi_1(X) \otimes 1)E_2 \tag{5.13}$$

*for all* $X \in \mathcal{X}$,
(4) *There is a Hilbert space* $\mathcal{K}$ *such that* $\pi_1 \otimes 1_{\mathbb{B}(\mathcal{K})} \simeq \pi_2 \otimes 1_{\mathbb{B}(\mathcal{K})}$.

It is also shown that $\pi_1 \approx \pi_2$ if and only if the ultraweak extensions $\widetilde{\pi}_1$ and $\widetilde{\pi}_2$ of $\pi_1$ and $\pi_2$ to $\mathcal{X}^{**}$ are physically equivalent [16, Chap. II, Theorem 12]. By the proof of the above theorem, we can show a similar theorem for quasi-containment of representations of C*-algebras.

**Theorem 12** ([14, Chap. III, Theorem 2.7]) *Let* $\mathcal{M}$ *be a von Neumann algebra on a Hilbert space* $\mathcal{H}$ *and* $\mathcal{V}$ *a closed subspace of* $\mathcal{M}_*$.
$(i)$ *If there exists a subset* $\mathcal{A}$ *of* $\mathcal{M}$*, whose linear combinations are ultraweakly dense in* $\mathcal{M}$*, such that* $A\mathcal{V} \subset \mathcal{V}$ *(resp.* $\mathcal{V}A \subset \mathcal{V}$*) for every* $A \in \mathcal{A}$*, then* $\mathcal{V}$ *is left (resp. right) invariant.*
$(ii)$ *There exists a bijection* $\mathcal{V} \leftrightarrow \mathcal{I}$ *between the closed left (resp. right) invariant subspaces* $\mathcal{V}$ *of* $\mathcal{M}_*$ *and the ultraweakly closed right (resp. left) ideals* $\mathcal{I}$ *of* $\mathcal{M}$ *determined by*

$$\mathcal{V}^0 = \mathcal{I} \quad \text{and} \quad \mathcal{I}^0 = \mathcal{V}, \tag{5.14}$$

*where* $\mathcal{V}^0$ *and* $\mathcal{I}^0$ *mean the polar of* $\mathcal{V}$ *and* $\mathcal{I}$ *in* $\mathcal{M}$ *and* $\mathcal{M}_*$*, respectively.*
$(iii)$ *Every left (resp. right) invariant subspace* $\mathcal{V}$ *is of the form*

$$\mathcal{V} = \mathcal{M}_*E \quad (\text{resp. } \mathcal{V} = E\mathcal{M}_*) \tag{5.15}$$

*for some projection* $E$ *of* $\mathcal{M}$*. Furthermore,* $\mathcal{V}$ *is two-sided invariant if and only if* $E$ *is an element of the center* $\mathcal{Z}(\mathcal{M})$ *of* $\mathcal{M}$.

Let $\pi$ be a representation of $\mathcal{X}$ on a Hilbert space $\mathcal{H}$. $\mathcal{V}(\pi)$ denotes the set $\{\omega \circ \pi \mid \omega \in \pi(\mathcal{X})''_*\}$, which is a two-side invariant subspace of $\mathcal{X}^*$. Then $C_\pi$ denotes the projection of $\mathcal{Z}(\mathcal{X}^{**})$ such that $\mathcal{V}(\pi) = \mathcal{X}^*C_\pi$. In particular, we have $\mathcal{S}_\pi(\mathcal{X}) = \{\varphi \in \mathcal{S}(\mathcal{X}) \mid C_\pi\varphi = \varphi\}$. Thus we have the following theorem.

**Theorem 13** ([14, Chap. III, Proposition 2.12]) *Let $\mathcal{X}$ be a C*-algebra. Let $\pi_1$ and $\pi_2$ be representations of $\mathcal{X}$ on Hilbert spaces $\mathcal{H}_1$ and $\mathcal{H}_2$, respectively. The following conditions are equivalent:*
(1) $\pi_1 \approx \pi_2$.
(2) $\mathcal{V}(\pi_1) = \mathcal{V}(\pi_2)$.
(3) $C_{\pi_1} = C_{\pi_2}$.

We shall define a notion called disjointness, which is a counterpart to quasi-equivalence.

**Definition 11** (*Disjointness*) Let $\mathcal{X}$ be a C*-algebra. Let $\pi_1$ and $\pi_2$ be representations of $\mathcal{X}$ on Hilbert spaces $\mathcal{H}_1$ and $\mathcal{H}_2$, respectively. $\pi_1$ and $\pi_2$ are said to be (mutually) disjoint, denoted by $\pi_1 \,\mathring{\mathrm{d}}\, \pi_2$, if there exists no non-zero $U \in \mathbb{B}(\mathcal{H}_1, \mathcal{H}_2)$ such that $U\pi_1(X) = \pi_2(X)U$ for all $X \in \mathcal{X}$. Two positive linear functional $\omega_1$ and $\omega_2$ on $\mathcal{X}$ are (mutually) disjoint if so are $\pi_{\omega_1}$ and $\pi_{\omega_2}$.

As in the case of quasi-equivalence, the following theorem holds for disjointness:

**Theorem 14** ([14, Proposition 2.12]) *Let $\mathcal{X}$ be a C*-algebra. Let $\pi_1$ and $\pi_2$ be representations of $\mathcal{X}$ on Hilbert spaces $\mathcal{H}_1$ and $\mathcal{H}_2$, respectively. The following conditions are equivalent:*
(1) $\pi_1 \,\mathring{\mathrm{d}}\, \pi_2$;
(2) *No $\pi_1$-normal state is $\pi_2$-normal, and vice versa, i.e., $\mathcal{S}_{\pi_1}(\mathcal{X}) \cap \mathcal{S}_{\pi_2}(\mathcal{X}) = \emptyset$.*

**Definition 12** (*Factor state*) A state $\omega$ on $\mathcal{X}$ is factor if $\mathcal{Z}_\omega(\mathcal{X}) = \pi_\omega(\mathcal{X})'' \cap \pi_\omega(\mathcal{X})' = \mathbb{C}1$. $\mathcal{S}_f(\mathcal{X})$ denotes the set of factor states on $\mathcal{X}$.

For factor states, the following theorem of alternatives holds.

**Theorem 15** (Dixmier [10, Corollary 5.3.6]) *Two different factor states on a C*-algebra $\mathcal{X}$ are either quasi-equivalent or disjoint.*

**Definition 13** Let $\omega_1$ and $\omega_2$ be positive linear functionals on $\mathcal{X}$. We say that $\omega_1$ and $\omega_2$ are (mutually) orthogonal if there exists no positive linear functional $\omega$ on $\mathcal{X}$ such that

$$\omega(X) \leq \omega_1(X) \quad \text{and} \quad \omega(X) \leq \omega_2(X) \tag{5.16}$$

for all $X \in \mathcal{X}_+$.

**Proposition 4** ([7, Lemma 4.1.19]) *Let $\omega_1$ and $\omega_2$ be positive linear functionals on $\mathcal{X}$. We then put $\omega = \omega_1 + \omega_2$. The following conditions are equivalent:*
(1) *$\omega_1$ and $\omega_2$ are mutually orthogonal.*
(2) *There exists a projection $P$ of $\pi_\omega(\mathcal{X})'$ such that*

$$\omega_1(X) = \langle \Omega_\omega | P\pi_\omega(X)\Omega_\omega \rangle \quad \text{and} \quad \omega_2(X) = \langle \Omega_\omega | (1-P)\pi_\omega(X)\Omega_\omega \rangle \tag{5.17}$$

*for all $X \in \mathcal{X}$.*
(3) $(\pi_\omega, \mathcal{H}_\omega, \Omega_\omega) = (\pi_{\omega_1} \oplus \pi_{\omega_2}, \mathcal{H}_{\omega_1} \oplus \mathcal{H}_{\omega_2}, \Omega_{\omega_1} \oplus \Omega_{\omega_2})$.

Let $\mu$ be a regular Borel measure on $\mathcal{S}(\mathcal{X})$. For every $\Delta \in \mathcal{B}(\mathcal{S}(\mathcal{X}))$, we define a positive linear functional $b(\Delta;\mu)$ on $\mathcal{X}$ by

$$b(\Delta;\mu)(X) = \int_{\Delta} \rho(X)\, d\mu(\rho) \tag{5.18}$$

for all $X \in \mathcal{X}$. $b(\mu)$ denotes $b(\mathcal{S}(\mathcal{X});\mu)$ and is called the barycenter of $\mu$.

**Definition 14** ([7, *Definition 4.1.20*]) Let $\mu$ be positive regular Borel measure on $\mathcal{S}(\mathcal{X})$. $\mu$ is called an orthogonal measure on $\mathcal{S}(\mathcal{X})$ if $b(\Delta;\mu)$ and $b(\mathcal{S}(\mathcal{X})\backslash\Delta;\mu)$ are mutually orthogonal for all $\Delta \in \mathcal{B}(\mathcal{S}(\mathcal{X}))$. An orthogonal measure $\mu$ is called an orthogonal measure of a positive linear functional $\omega$ on $\mathcal{X}$ if $b(\mu) = \omega$.

For every $X \in \mathcal{X}$, we define $\hat{X} \in C(\mathcal{S}(\mathcal{X}))$ by

$$\hat{X}(\omega) = \omega(X) \tag{5.19}$$

for all $\omega \in \mathcal{S}(\mathcal{X})$.

**Theorem 16** ([7, Theorem 4.1.25]) *Let $\mathcal{X}$ be a $C^*$-algebra and $\omega$ a state on $\mathcal{X}$. There exists a one-to-one correspondence among the following three sets:*
*(1) The set of abelian von Neuman subalgebras $\mathcal{B}$ of the commutant $\pi_\omega(\mathcal{X})'$ of $\pi_\omega(\mathcal{X})''$;*
*(2) The set of orthogonal measures $\mu$ of $\omega$;*
*(3) The set of projections $P$ on $\mathcal{H}_\omega$ such that $P\pi_\omega(\mathcal{X})P \subseteq (P\pi_\omega(\mathcal{X})P)'$.*
*It $\mathcal{B}$, $\mu$ and $P$ are in the above one-to-one correspondence, then the following relations hold:*
*(i) There exists the $W^*$-isomorphism $\kappa_\mu : L^\infty(\mathcal{S}(\mathcal{X}),\mu) \to \mathcal{B}$ defined by*

$$\langle \Omega_\omega | \kappa_\mu(f)\pi_\omega(X)\Omega_\omega\rangle = \int_{\mathcal{S}(\mathcal{X})} f(\rho)\hat{X}(\rho)\, d\mu(\rho) \tag{5.20}$$

*for all $X \in \mathcal{X}$ and $f \in L^\infty(\mathcal{S}(\mathcal{X}),\mu)$.*
*(ii)*

$$\int_{\mathcal{S}(\mathcal{X})} \hat{X}_1(\rho)\cdots\hat{X}_n(\rho)\, d\mu(\rho) = \langle \Omega_\omega | \pi_\omega(X_1)P\cdots P\pi_\omega(X_n)\Omega_\omega\rangle \tag{5.21}$$

*for all $n \in \mathbb{N}$ and $X_1,\ldots,X_n \in \mathcal{X}$.*
*(iii) $\mathcal{B} = (\pi_\omega(\mathcal{X}) \cup \{P\})'$.*
*(iv) $P = [\mathcal{B}\Omega_\omega]$, where $[\mathcal{B}\Omega_\omega]$ is a projection on $\mathcal{H}_\omega$ whose range is the closed linear subspace of $\mathcal{H}_\omega$ spanned by $\mathcal{B}\Omega_\omega$.*

**Corollary 4** *Let $\mathcal{X}$ be a $C^*$-algebra and $\omega$ a state on $\mathcal{X}$. There exists a one-to-one correspondence between the following two sets:*
*(1) The set of von Neuman subalgebras $\mathcal{B}$ of the center $\mathfrak{Z}_\omega(\mathcal{X})$ of $\pi_\omega(\mathcal{X})''$;*

(2) *The set of subcentral measures* $\mu$ *of* $\omega$*, i.e.,* $b(\Delta; \mu)$ *and* $b(\mathcal{S}(\mathcal{X})\backslash\Delta; \mu)$ *are mutually disjoint for all* $\Delta \in \mathcal{B}(\mathcal{S}(\mathcal{X}))$*;*
*The above one-to-one correspondence is given by the* $W^*$*-isomorphism* $\kappa_\mu : L^\infty(\mathcal{S}(\mathcal{X}), \mu) \to \mathcal{B}$ *defined by*

$$\langle \Omega_\omega | \kappa_\mu(f) \pi_\omega(X) \Omega_\omega \rangle = \int_{\mathcal{S}(\mathcal{X})} f(\rho) \hat{X}(\rho) \, d\mu(\rho) \tag{5.22}$$

*for all* $X \in \mathcal{X}$ *and* $f \in L^\infty(\mathcal{S}(\mathcal{X}), \mu)$*.*

$\mu_\omega$ denotes the barycentric measure $\mu$ of $\omega$ corresponding to the center $\mathfrak{Z}_\omega(\mathcal{X})$ of $\pi_\omega(\mathcal{X})''$, and is called the central measure of $\omega$. $\mu_{\omega,\mathcal{B}}$ denotes the barycentric measure $\mu$ of $\omega$ corresponding to a von Neumann subalgebra $\mathcal{B}$ of $\pi_\omega(\mathcal{X})'$.

**Definition 15** Let $\mu$ be a Borel measure on a locally compact Hausdorff space $S$. Let $A$ be a subset of $S$. We say that $\mu$ is pseudosupported by $A$ if $\mu(\Delta) = 0$ for all $\Delta \in \mathcal{B}(S)$ such that $A \cap \Delta = \emptyset$.

**Theorem 17** ([7, Theorem 4.2.10]) *Let* $\mathcal{X}$ *be a* $C^*$*-algebra and* $\omega$ *a state on* $\mathcal{X}$*. The central measure* $\mu_\omega$ *of* $\omega$ *is pseudosupported by* $\mathcal{S}_f(\mathcal{X})$*. If* $\mathcal{X}$ *is separable, then* $\mu_\omega$ *is supported by* $\mathcal{S}_f(\mathcal{X})$*.*

## 5.2.2 (C*-)Algebraic Quantum Theory

### 5.2.2.1 Von Neumann's Axiomatic System

To begin with, we shall review von Neumann's axiomatic system of quantum mechanics [17]. The axiomatic system of algebraic quantum theory introduced in this section is of course its algebraic modification.

**Axiom 1** A physical system **S** is specified by a Hilbert space $\mathcal{H}$. That is to say, observables of **S** correspond to self-adjoint operators $A$ on $\mathcal{H}$, and physical situtations and experimental settings of **S** are described by density operators $\rho$ on $\mathcal{H}$.

**Axiom 2** (*Born statistical formula*) Let $A$ be an observable of **S** and $\rho$ a state of **S**. For all $\Delta \in \mathcal{F}$, the probability $\Pr\{A \in \Delta \| \rho\}$ that values of $A$ contained in $\Delta \in \mathcal{B}(\mathbb{R})$ emerges in $\rho$ is defined by

$$\Pr\{A \in \Delta \| \rho\} = \mathrm{Tr}[E^A(\Delta)\rho], \tag{5.23}$$

where $E^A$ denotes the spectral measure of $A$.

**Axiom 3** Dynamical changes of the system **S** are described by adjoint actions $AdU$ on self-adjoint operators (or density operators) of unitary operators $U$ on $\mathcal{H}$. In particular, are described by unitary representations $\{U_t\}_{t\in\mathbb{R}}$ on $\mathcal{H}$.

**Axiom 4** Let $\mathbf{S}_1$ and $\mathbf{S}_2$ be physical systems described by Hilbert spaces $\mathcal{H}_1$ and $\mathcal{H}_2$, respectively. Then the composite system $\mathbf{S}_1 + \mathbf{S}_2$ of $\mathbf{S}_1$ and $\mathbf{S}_2$ is described by the tensor product Hilbert space $\mathcal{H}_1 \otimes \mathcal{H}_2$. Furthermore, observables $A$ of $\mathbf{S}_1$ are identified with $A \otimes 1$ of $\mathbf{S}_1 + \mathbf{S}_2$, and observables $B$ of $\mathbf{S}_2$ are identified with $1 \otimes B$ of $\mathbf{S}_1 + \mathbf{S}_2$.

It is clear that this axiomatic system is very useful for quantum systems with finite degrees of freedom. On the other hand, it cannot systematically characterize those with infinite degrees of freedom. Another disadvantage of the axiomatic system is that the meaning of the concept of state is made through Axiom 2. Originally, it must also take into account that every state defines not only probability distributions of observables but also correlation functions. Thus we need another axiomatic approach to quantum systems as a working hypothesis. (C*-)Algebraic quantum theory is one of the attempts.

#### 5.2.2.2 An Axiomatic System of Algebraic Quantum Theory

The axiomatic system of algebraic quantum theory is very simple except for the description of the dynamics. Algebraic quantum theory is based on algebraic (or noncommutative, quantum) probability theory. We refer the reader to [18] for details on algebraic probability theory. The following is the most important axiom in (algebraic) quantum theory as a working hypothesis.

**Axiom 5** (*Observables and states* [19, 20]) All the statistical aspects of a physical system are registered in a C*-probability space $(\mathcal{X}, \omega)$. Observables are described by self-adjoint elements of $\mathcal{X}$. On the other hand, $\omega$ statistically corresponds to a physical situation (or an experimental setting).

Every state functions as an expectation functional for all observables of the system. At the same time, its has macroscopic aspects defined by the quasi-equivalence of representations.

**Definition 16** (*Sector* [5]) A sector of $\mathcal{X}$, denoted by $[\omega]$, is a quasi-equivalence class of a factor state $\omega$ on $\mathcal{X}$. $\widehat{\mathcal{X}}$ denotes the set of sectors of $\mathcal{X}$, i.e., $\widehat{\mathcal{X}} = \mathcal{S}_f(\mathcal{X})/\approx$.

By Theorem 17, the central measure $\mu_\omega$ of a state $\omega$ is pseudosupported by $\mathcal{S}_f(\mathcal{X})$. By Corollary 4, there then exists a $W^*$-isomorphism $\kappa_\omega : L^\infty(\mathcal{S}(\mathcal{X}), \mu_\omega) \to \mathcal{Z}_\omega(\mathcal{X})$ such that

$$\langle \Omega_\omega | \kappa_\omega(f) \pi_\omega(X) \Omega_\omega \rangle = \int_{\mathcal{S}(\mathcal{X})} f(\rho) \hat{X}(\rho) \, d\mu_\omega(\rho) \tag{5.24}$$

for all $X \in \mathcal{X}$ and $f \in L^\infty(\mathcal{S}(\mathcal{X}), \mu_\omega)$. Thus it is natural to consider that we can distinguish sectors emerged in a state $\omega$ by using observables in the center $\mathcal{Z}_\omega(\mathcal{X})$ of $\pi_\omega(\mathcal{X})''$. It is valid to adopt the following axiom in this context.

**Axiom 6** (*Sector as event* [19, 20]) For a state $\omega$ and a Borel set $\Delta$ of $\mathcal{S}(\mathcal{X})$, $\mu_\omega(\Delta)$ gives the probability that sectors as elementary events belong to $\Delta$ under the situation described by $\omega$. When available observables are restricted, the coarse-grained probability is given by $\mu_{\omega,\mathcal{B}}(\Delta)$ for some von Neumann subalgebra $\mathcal{B}$ of $\mathcal{Z}_\omega(\mathcal{X})$.

The axiom states that sectors give macroscopic classical levels in quantum systems and that we can describe macrosopic quantities of quantum systems by measure-theoretical probability theory. In particular, each sector corresponds to an elementary event. Futhermore, the Born statistical formula in Axiom 2 follows from this axiom by considering measuring processes [19, 20].

Different from the above two axioms, it is now difficult to adopt axioms of dynamics and composite systems. This is because various aspects we should consider are unsettled.

### 5.2.3 Sector and Symmetry

We discuss the relation between sector theory and symmetry herein. Symmetry has been played the crucial role in physics. C*-algebraic quantum theory enables us to flexibly describe symmetry in quantum theory. In particular, we can systematically formulate physics and mathematics of symmetry breaking by using operator algebras.

Before discussing symmetry breaking, we shall define invariance for symmetry in a given state.

**Definition 17** Let $\mathcal{X}$ be a C*-algebra, $G$ a locally compact group, $\alpha$ a strongly continuous action of $G$ on $\mathcal{X}$ and $\omega$ a state on $\mathcal{X}$. We say that $\omega$ is $(G, \alpha)$-invariant (or $G$-invariant simply) if

$$\omega(\alpha_g(X)) = \omega(X) \tag{5.25}$$

for all $X \in \mathcal{X}$ and $g \in G$.

For every $G$-invariant state $\omega$, the following representation theorem holds:

**Theorem 18** *If $\omega$ is $(G, \alpha)$-invariant, then there exists a unitary representation $U_\omega : G \to \mathbb{B}(\mathcal{H}_\omega)$ such that*

$$\pi_\omega(\alpha_g(X)) = U_\omega(g)^* \pi_\omega(X) U_\omega(g), \tag{5.26}$$

$$U_\omega(g)\Omega_\omega = \Omega_\omega \tag{5.27}$$

*for all $X \in \mathcal{X}$ and $g \in G$.*

Thus the $G$-invariance of states enables us to use the theory of unitary representations of locally compact Hausdorff groups. It is, however, not sufficient to systematically formulate symmetry breaking in quantum theory, and is replaced by the $G$-quasi-invariance of states. From our point of view, there is no need to stick to the $G$-invariance of states since it is more restrictive than the $G$-quasi-invariance.

From now on, we shall discuss symmetry breaking by two methods. The difference between the two methods is whether the state of interest is prepared before or after "eventualization".

Let $\mathcal{X}$ be a C*-algebra, $G$ a locally compact group, $\alpha$ a strongly continuous action of $G$ on $\mathcal{X}$. We define the action $\alpha$ of $G$ on $\mathcal{X}^*$ by

$$\alpha_g\omega(X) = \omega(\alpha_{g^{-1}}(X)) \tag{5.28}$$

for all $\omega \in \mathcal{X}$, $g \in G$ and $X \in \mathcal{X}$.

#### 5.2.3.1 First Look at Symmetry Breaking

To begin with, we shall observe the following proposition.

**Proposition 5** *Let $\mathcal{X}$ be a C*-algebra, G a separable locally compact Hausdorff group and $\alpha$ a strongly continuous action of G on $\mathcal{X}$. Let $\pi$ be a representation of $\mathcal{X}$ on a separable Hilbert space $\mathcal{H}$. Then the following conditions are equivalent:*
(1) *For every $g \in G$, $\pi \approx \pi \circ \alpha_g$.*
(2) *There exists an ultraweakly continuous action $\tilde{\alpha}$ of G on $\pi(\mathcal{X})''$ such that $\pi \circ \alpha_g = \tilde{\alpha}_g \circ \pi$ for all $g \in G$.*

**Definition 18** (*G-quasi-invariance*) Let $\mathcal{X}$ be a C*-algebra, $G$ a separable locally compact Hausdorff group and $\alpha$ a strongly continuous action of $G$ on $\mathcal{X}$.
(1) A representation $\pi$ of $\mathcal{X}$ on a separable Hilbert space $\mathcal{H}$ is said to be $(G, \alpha)$-quasi-invariant (or $G$-quasi-invariant simply) if it satisfies the equivalent conditions in Proposition 5.
(2) A state $\omega$ on $\mathcal{X}$ is said to be $(G, \alpha)$-quasi-invariant (or $G$-quasi-invariant simply) if $\alpha_g\omega \approx \omega$ for all $g \in G$.

For all $\omega \in E_{\mathcal{X}}$ and $\Delta \in \mathcal{B}(\mathcal{S}(\mathcal{X}))$ such that $\mu_\omega(\Delta) \neq 0$, we define $\omega_\Delta$ by

$$\omega_\Delta = \frac{1}{\mu_\omega(\Delta)} \int_\Delta \rho \, d\mu_\omega(\rho). \tag{5.29}$$

Here we define both unbroken symmetry and broken symmetry, which are the state version of the definition by Ojima [3–5].

**Definition 19** (*Unbroken symmetry and broken symmetry*)

(1) The strongly continuous action $\alpha$ of $G$ on $\mathcal{X}$ is said to be unbroken in $\omega$ if $\omega_\Delta \approx \alpha_g\omega_\Delta$ for all $g \in G$ and $\Delta \in \mathcal{B}(\mathcal{S}(\mathcal{X}))$ such that $\mu_\omega(\Delta) \neq 0$.
(2) The strongly continuous action $\alpha$ of $G$ on $\mathcal{X}$ is said to be broken in $\omega$ if it is not unbroken in $\omega$.

By the definition, we say that an action $\alpha$ of $G$ is unbroken in $\omega$ if all sectors involved in $\omega$ is $(G, \alpha)$-quasi-invariant. We now consider the theory of symmetry breaking in situations where the whole space of order parameters is specified. Suppose that

1. The GNS representation of $\omega$ is separable, i.e., $\mathcal{H}_\omega$ is separable.
2. The action $\alpha$ of $G$ on $\mathcal{X}$ is broken in $\omega$,
3. $\omega \approx \alpha_g \omega$ for all $g \in G$.

By Proposition 5, there exists an ultraweakly continuous action $\tilde{\alpha}$ of $G$ on $\pi_\omega(\mathcal{X})''$ such that $\pi_\omega \circ \alpha_g = \tilde{\alpha}_g \circ \pi_\omega$ for all $g \in G$. $\mathfrak{Z}_\omega(\mathcal{X})$ We define an ultraweakly continuous action $\widehat{\alpha}$ of $G$ on $\mathfrak{Z}_\omega(\mathcal{X})$ by

$$\widehat{\alpha}_g\,(Z) = \tilde{\alpha}_g(Z) \tag{5.30}$$

for all $Z \in \mathfrak{Z}_\omega(\mathcal{X})$. We now define the situation that the action $\alpha$ of $G$ is broken in $\omega$ and the action $\alpha|_H$ of $H$ is unbroken in $\omega$. We say that the action $\alpha$ of $G$ is broken in $\omega$ and the action $\alpha|_H$ of $H$ is unbroken in $\omega$ if there exists an isomorphism between $W^*$-dynamical systems

$$(\mathfrak{Z}_\omega(\mathcal{X}), G, \widehat{\alpha}) \cong (L^\infty(H\backslash G, \mu), G, \tau), \tag{5.31}$$

where $\mu$ is a $G$-quasi-invariant regular Borel measure on $H\backslash G$ and $\tau$ is the action of $G$ on $L^\infty(H\backslash G, \mu)$ defined by

$$\tau_g(f)(x) = f(xg) \tag{5.32}$$

for all $f \in L^\infty(H\backslash G, \mu)$, $g \in G$ and $x \in H\backslash G$.

We then assume that $G$ is a Lie group and $H$ is a closed Lie subgroup. $\mathfrak{g}$ and $\mathfrak{h}$ denote the Lie algebras of $G$ and $H$, respectively. For every $X \in \mathfrak{g}$, we define the derivative $\tilde{\delta}_X : \mathrm{dom}(\tilde{\delta}_X) \to \pi_\omega(\mathcal{X})''$ by

$$\mathrm{dom}(\tilde{\delta}_X) = \{A \in \pi_\omega(\mathcal{X})'' \mid \lim_{t\to 0} \frac{\tilde{\alpha}_{e^{tX}}(A) - A}{t} \text{ exists } \}, \tag{5.33}$$

$$\tilde{\delta}_X(A) = \lim_{t\to 0} \frac{\tilde{\alpha}_{e^{tX}}(A) - A}{t} \tag{5.34}$$

for all $A \in \mathrm{dom}(\tilde{\delta}_X)$.

Furthermore, for every $X \in \mathfrak{g}$, the restriction $\widehat{\delta}_X$ of $\tilde{\delta}_X$ to $\mathfrak{Z}_\omega(\mathcal{X})$ is defined by

$$\mathrm{dom}(\widehat{\delta}_X) = \mathrm{dom}(\tilde{\delta}_X) \cap \mathfrak{Z}_\omega(\mathcal{X}), \tag{5.35}$$

$$\widehat{\delta}_X\,(Z) = \tilde{\delta}_X(Z) \tag{5.36}$$

for all $Z \in \mathrm{dom}(\widehat{\delta}_X)$.

The mathematical tools presented here are very useful for the description of differential geometric aspects of quantum systems.

#### 5.2.3.2 Second Look at Symmetry Breaking

Here we consider the theory of symmetry breaking in situations where one point in the space of order parameters is specified. Let $G$ be a locally compact Hausdorff group and $H$ a closed subgroup of $G$. Let $\mathcal{X}$ be a C*-algebra, $\omega$ a factor state on $\mathcal{X}$ and $\beta$ a strongly continuous action of $H$ on $\mathcal{X}$. Suppose that the action $\beta$ of $H$ is unbroken in $\omega$. We would like to realize the situation where $G/H$ becomes the space of order parameters. At this time, one of the following two is assumed.

1. There exists a strongly continuous action $\alpha$ of $G$ on $\mathcal{X}$ such that $\alpha_h = \beta_h$ for all $h \in H$ and such that $\alpha_{g_1}\omega \,\delta\, \alpha_{g_2}\omega$ for every $(g_1, g_2) \in \{(g_1, g_2) \in G \times G \mid g_2^{-1}g_1 \notin H\}$.
2. There exists no strongly continuous action $\alpha$ of $G$ on $\mathcal{X}$ such that $\alpha_h = \beta_h$ for all $h \in H$ and such that $\alpha_{g_1}\omega \,\delta\, \alpha_{g_2}\omega$ for every $(g_1, g_2) \in \{(g_1, g_2) \in G \times G \mid g_2^{-1}g_1 \notin H\}$.

In the case 1, we can apply the method as mentioned before (First look at symmetry breaking) to the state $\omega_\mu$ defined by

$$\omega_\mu(X) = \int_{G/H} \alpha_{s(x)}\omega(X)\, d\mu(x) \tag{5.37}$$

for all $X \in \mathcal{X}$, where $\mu$ is a $G$-quasi-invariant regular Borel measure on $G/H$ and $s : G/H \to G$ is a measurable section. Thus the case 1 is equivalent to the situation that a $G$-quasi-equivalent state $\varphi$ is given, whose $W^*$-dynamical system $(\mathfrak{Z}_\varphi(\mathcal{X}), G, \widehat{\alpha})$ satisfies

$$(\mathfrak{Z}_\varphi(\mathcal{X}), G, \widehat{\alpha}) \cong (L^\infty(H\backslash G, \mu), G, \tau) \tag{5.38}$$

as $W^*$-dynamical system.

The case 2 is completely different from the case 1. It is no longer possible to deal with symmetry breaking due to the action of group $G$ on the specified observable algebra alone. Therefore, we need the notion of augmented algebra inspired by induced representations of groups. The introduction of augmented algebra enables us to discuss explicitly broken symmetry. In the physical context, only spontaneously broken symmetry is discussed. Originally, spontaneously broken symmetry and explicitly broken symmetry should be treated equally. On the other hand, the latter is a substitute which could only be discussed under appropriate concept settings. Therefore, we had to wait for investigations by Ojima [3–5].

## 5.3 An Introduction to Modern Measurement Theory

We introduce foundations of quantum measurement theory herein. We refer the reader to [21–23] for detailed introductory expositions of quantum measurement theory. From our point of view, quantum measurement theory is not physically difficult but mathematically a little complicated. Nonetheless, it gives us various inspiration. We cannot systematically understand quantum theory without the understanding for processes of measurement. If the reader touches this topic for the first time, knowing the overview is more important than anything else.

In this section, we assume that von Neumann algebras $\mathcal{M}$ are $\sigma$-finite. However, only in the case of $\mathcal{M} = \mathbb{B}(\mathcal{H})$, the von Neumann algebra of bounded operators on a Hilbert space $\mathcal{H}$, this assumption is ignored. The sentences in this section overlap somewhat with [24].

### *5.3.1 Quantum Measurement Theory Before Davies-Lewis*

The old quantum measurement theory is (implicitly) based on a postulate called the repeatability hypothesis.

**Axiom 7** (*Repeatability hypothesis* [17]) If a physical quantity is measured successively twice, then we get the same value at each time.

Under the hypothesis, von Neumann derived the following postulate for nondegenerate discrete observables:

**Axiom 8** (*von Neumnan–Lüders projection postulate*) Let $A$ be a discrete observable on a Hilbert space $\mathcal{H}$ and $\rho$ a density operator on $\mathcal{H}$. We consider the situation where $\rho$ is the initial state. The state after the measurement of $A$ when we get values of $A$ contained in $\Delta \in \mathcal{B}(\mathbb{R})$ is given by

$$\frac{\sum_{a\in\Delta} E^A(\{a\})\rho E^A(\{a\})}{\mathrm{Tr}[E^A(\Delta)]} \tag{5.39}$$

if $\mathrm{Tr}[E^A(\Delta)\rho] > 0$.

The mathematical essence of the above postulate was revealed by Nakamura and Umegaki [25]. After the study of Arveson [26], Davies and Lewis [27] proposed that we should abandon the repeatability hypothesis as a general principle and employ an operational approach to quantum measurement. References [27, 28] will be helpful for the detailed circumstances of this topic.

### *5.3.2 Modern Measurement Theory*

Davies and Lewis [27] introduced the notion of instruments (Definition 20) to develop quantum measurement theory independent of the reatability hypothesis. Their proposal is summarized as follows:

**Davies–Lewis proposal** *For every apparatus* $\mathbf{A}(\mathbf{x})$ *measuring* $\mathbf{S}$*, where* $\mathbf{x}$ *is the output variable of* $\mathbf{A}(\mathbf{x})$ *taking values in a measurable space* $(S, \mathcal{F})$*, there always exists an instrument* $\mathcal{I}$ *for* $(\mathcal{M}, S)$ *corresponding to* $\mathbf{A}(\mathbf{x})$ *in the following sense. For every input state* $\rho$*, the probability distribution* $\Pr\{\mathbf{x} \in \Delta \| \rho\}$ *of* $\mathbf{x}$ *in* $\rho$ *is given by*

$$\Pr\{\mathbf{x} \in \Delta \| \rho\} = \|\mathcal{I}(\Delta)\rho\| = \langle \mathcal{I}(\Delta)\rho, 1\rangle \tag{5.40}$$

*for all* $\Delta \in \mathcal{F}$*, and the state* $\rho_{\{\mathbf{x}\in\Delta\}}$ *after the measurement under the condition that* $\rho$ *is the prepared state and the outcome* $\mathbf{x} \in \Delta$ *is given by*

$$\rho_{\{\mathbf{x}\in\Delta\}} = \frac{\mathcal{I}(\Delta)\rho}{\|\mathcal{I}(\Delta)\rho\|} \tag{5.41}$$

*if* $\Pr\{\mathbf{x} \in \Delta \| \rho\} > 0$*, and* $\rho_{\{\mathbf{x}\in\Delta\}}$ *is indefinite if* $\Pr\{\mathbf{x} \in \Delta \| \rho\} = 0$.

In other words, they claimed that experimentally and statistically accessible ingredients via measurements by a given measuring apparatus should be specified by instruments. No epochal investigation of quantum measurement theory using instruments appeared soon after their study. However, the circumstance was entirely changed after the study of Ozawa [29–31]. He introduced the notion of completely positive (CP) instrument (Definition 21) and that of measuring process (Definition 22). The latter is nothing but quantum mechanical models of measuring appratuses. Furthermore, he established the one-to-one correspondence between CP instruments and "statistical equivalence classes" of measuring processes (Theorem 19). The following criterion is widely accepted now.

**Davies–Lewis–Ozawa criterion** *For every apparatus* $\mathbf{A}(\mathbf{x})$ *measuring* $\mathbf{S}$*, where* $\mathbf{x}$ *is the output variable of* $\mathbf{A}(\mathbf{x})$ *taking values in a measurable space* $(S, \mathcal{F})$*, there always exists a CP instrument* $\mathcal{I}$ *for* $(\mathcal{M}, S)$ *corresponding to* $\mathbf{A}(\mathbf{x})$ *in the sense of the Davies-Lewis proposal, i.e., for every input state* $\rho$ *and outcome* $\Delta \in \mathcal{F}$ *both the probability distribution* $\Pr\{\mathbf{x} \in \Delta \| \rho\}$ *of* $\mathbf{x}$ *and the state* $\rho_{\{\mathbf{x}\in\Delta\}}$ *after the measurement are obtained from* $\mathcal{I}$.

Our purpose of this section is to introduce the theory of quantum measurement after the study of the author with the collaborator, Ozawa, that opened up quantum measurement theory of quantum systems with infinite degrees of freedom including quantum fields. We shall treat the following four topics in the rest of this section:

1. Completely positive instruments
2. Repeatability and CP instruments without the NEP
3. Measurement theory in the Heisenberg picture
4. Approximate realization of CP instruments by measuring processes.

#### 5.3.2.1 Completely Positive Instruments

Let $\mathbf{S}$ be a system whose observables and states are described by self-adjoint operators affiliated to a von Neumann algebra $\mathcal{M}$ on a Hilbert space $\mathcal{H}$ and by normal states on

$\mathcal{M}$, respectively. $\mathcal{M}_*$ denotes the predual of $\mathcal{M}$, i.e., the set of ultraweakly continuous linear functionals on $\mathcal{M}$, $\mathcal{S}_n(\mathcal{M})$ does the set of normal states on $\mathcal{M}$ and $P(\mathcal{M}_*)$ does the set of positive linear maps on $\mathcal{M}_*$.

**Definition 20** (*Instrument* [27, p. 243, ll.21–26]) Let $\mathcal{M}$ be a von Neumann algebra on a Hilbert space $\mathcal{H}$ and $(S, \mathcal{F})$ a measurable space. A map $\mathcal{I} : \mathcal{F} \to P(\mathcal{M}_*)$ is called an instrument for $(\mathcal{M}, S)$ if it satisfies the following conditions:
(1) $\langle \mathcal{I}(S)\rho, 1\rangle = \langle \rho, 1\rangle$ for all $\rho \in \mathcal{M}_*$;
(2) For every $M \in \mathcal{M}$, $\rho \in \mathcal{M}_*$ and mutually disjoint sequence $\{\Delta_j\}$ of $\mathcal{F}$,

$$\langle \mathcal{I}(\cup_j \Delta_j)\rho, M\rangle = \sum_j \langle \mathcal{I}(\Delta_j)\rho, M\rangle, \tag{5.42}$$

where $\langle \cdot, \cdot\rangle$ denotes the duality pairing of $\mathcal{M}_*$ and $\mathcal{M}$.

We define the dual map $\mathcal{I}^*$ of an instrument $\mathcal{I}$ by $\langle \rho, \mathcal{I}^*(\Delta)M\rangle = \langle \mathcal{I}(\Delta)\rho, M\rangle$ and use the notation $\mathcal{I}(M, \Delta) = \mathcal{I}^*(\Delta)M$ for all $\Delta \in \mathcal{F}$ and $M \in \mathcal{M}$.

By the definition, it is seen that an instrument describes the weighted state changes caused by the measurement. The dual map $\mathcal{I} : \mathcal{M} \times \mathcal{F} \to \mathcal{M}$ of an instrument $\mathcal{I}$ for $(\mathcal{M}, S)$ is characterized by the following conditions:
$(i)$ For every $\Delta \in \mathcal{F}$, the map $\mathcal{M} \ni M \mapsto \mathcal{I}(M, \Delta) \in \mathcal{M}$ is a normal positive linear map of $\mathcal{M}$;
$(ii)$ $\mathcal{I}(1, S) = 1$;
$(iii)$ For every $M \in \mathcal{M}$, $\rho \in \mathcal{M}_*$ and mutually disjoint sequence $\{\Delta_j\}$ of $\mathcal{F}$,

$$\langle \rho, \mathcal{I}(M, \cup_j \Delta_j)\rangle = \sum_j \langle \rho, \mathcal{I}(M, \Delta_j)\rangle. \tag{5.43}$$

Since a map $\mathcal{I} : \mathcal{M} \times \mathcal{F} \to \mathcal{M}$ satisfying the above conditions is always the dual map of an instrument for $(\mathcal{M}, S)$, we also call the map $\mathcal{I}$ an instrument for $(\mathcal{M}, S)$.

**Definition 21** (*CP instrument* [29, 30]) An instrument $\mathcal{I}$ for $(\mathcal{M}, S)$ is said to be *completely positive* if $\mathcal{I}(\Delta)$ (or $\mathcal{I}(\cdot, \Delta)$, equivalently) is completely positive for every $\Delta \in \mathcal{F}$. $\mathrm{CPInst}(\mathcal{M}, S)$ denotes the set of CP instruments for $(\mathcal{M}, S)$.

The notion of measuring process is defined as a quantum mechanical modeling of an apparatus as a physical system, of the meter of the apparatus, and of the measuring interaction between the system and the apparatus. Let $\mathcal{H}$ and $\mathcal{K}$ be Hilbert spaces. $\mathbb{B}(\mathcal{H})$ denotes the set of bounded linear operators on $\mathcal{H}$ and by $\mathbb{B}(\mathcal{H}, \mathcal{K})$ the set of bounded linear operators of $\mathcal{H}$ to $\mathcal{K}$. Let $\mathcal{M}$ and $\mathcal{N}$ be von Neumann algebras on $\mathcal{H}$ and $\mathcal{K}$, respectively. $\mathcal{M}\overline{\otimes}\mathcal{N}$ denotes the $W^*$-tensor product of $\mathcal{M}$ and $\mathcal{N}$. For every $\sigma \in \mathcal{N}_*$, a linear map $\mathrm{id} \otimes \sigma : \mathcal{M}\overline{\otimes}\mathcal{N} \to \mathcal{M}$ is defined by $\langle \rho, (\mathrm{id} \otimes \sigma)X\rangle = \langle \rho \otimes \sigma, X\rangle$ for all $X \in \mathcal{M}\overline{\otimes}\mathcal{N}$ and $\rho \in \mathcal{M}_*$. The following is the mathematical definition of measuring processes:

**Definition 22** (*Measuring process* [21, Definition 3.2]) A measuring process for $(\mathcal{M}, S)$ is a 4-tuple $\mathbb{M} = (\mathcal{K}, \sigma, E, U)$ of a Hilbert space $\mathcal{K}$, a normal state $\sigma$ on $\mathbb{B}(\mathcal{K})$, a PVM $E : \mathcal{F} \to \mathbb{B}(\mathcal{K})$ and a unitary operator $U$ on $\mathcal{H} \otimes \mathcal{K}$ satisfying

$$(\mathrm{id} \otimes \sigma)[U^*(M \otimes E(\Delta))U] \in \mathcal{M} \tag{5.44}$$

for all $M \in \mathcal{M}$ and $\Delta \in \mathcal{F}$.

Let $\mathbb{M} = (\mathcal{K}, \sigma, E, U)$ be a measuring process for $(\mathcal{M}, S)$. Then a CP instrument $\mathcal{I}_{\mathbb{M}}$ for $(\mathcal{M}, S)$ is defined by

$$\mathcal{I}_{\mathbb{M}}(M, \Delta) = (\mathrm{id} \otimes \sigma)[U^*(M \otimes E(\Delta))U] \tag{5.45}$$

for $M \in \mathcal{M}$ and $\Delta \in \mathcal{F}$. The most important example of measuring processes is a von Neumann model, which is the starting point of quantum mechanical modelings of appratuses [17, 32].

*Example 7* $(L^2(\mathbb{R}), \omega_\alpha, E^Q, e^{-iA\otimes P/\hbar})$ of measurement of an observable $A$, a self-adjoint operator affiliated with $\mathcal{M}$, where $\alpha$ is a unit vector of $L^2(\mathbb{R})$, $\omega_\alpha$ is defined by $\omega_\alpha(M) = \langle \alpha | M \alpha \rangle$ for all $M \in \mathbb{B}(L^2(\mathbb{R}))$, and $Q = \int_{\mathbb{R}} q \, dE^Q(q)$ and $P$ are self-adjoint operators defined on dense linear subspaces of $L^2(\mathbb{R})$ such that $\overline{[Q, P]} = i\hbar 1$. Here, we $E^X$ denotes the spectral measure of a self-adjoint operator $X$ densely defined on a Hilbert space.

**Definition 23** Two measuring processes $\mathbb{M}_1$ and $\mathbb{M}_2$ for $(\mathcal{M}, S)$ are *statistically equivalent* if they define an identical instrument, i.e., $\mathcal{I}_{\mathbb{M}_1} = \mathcal{I}_{\mathbb{M}_2}$.

In the case of $\mathcal{M} = \mathbb{B}(\mathcal{H})$, the following theorem, a unitary dilation theorem of CP instruments for $(\mathbb{B}(\mathcal{H}), S)$, is known to hold.

**Theorem 19** ([29], [30, Theorem 5.1], [21, Theorem 3.6]) *For every CP instrument $\mathcal{I}$ for $(\mathbb{B}(\mathcal{H}), S)$, there uniquely exists a statistical equivalence class of measuring processes $\mathbb{M} = (\mathcal{K}, \sigma, E, U)$ for $(\mathbb{B}(\mathcal{H}), S)$ such that $\mathcal{I}(M, \Delta) = \mathcal{I}_{\mathbb{M}}(M, \Delta)$ for all $M \in \mathbb{B}(\mathcal{H})$ and $\Delta \in \mathcal{F}$. Conversely, every statistical equivalence class of measuring processes for $(\mathbb{B}(\mathcal{H}), S)$ defines a unique CP instrument $\mathcal{I}$ for $(\mathbb{B}(\mathcal{H}), S)$.*

A generalization of this theorem to arbitrary von Neumann algebras is shown to hold not for all CP instruments but for those with the normal extension property (NEP) in [21] (see Theorem 20).

$L^\infty(S, \mathcal{I})$ denotes the $W^*$-algebra of essentially bounded $\mathcal{I}$-measurable functions on $S$ modulo the $\mathcal{I}$-negligible functions.

**Definition 24** (*Normal extension property* [21, Definition 3.3]) Let $\mathcal{M}$ be a von Neumann algebra on a Hilbert space $\mathcal{H}$ and $(S, \mathcal{F})$ a measurable space. Let $\mathcal{I}$ be a CP instrument for $(\mathcal{M}, S)$ and $\Psi_{\mathcal{I}} : \mathcal{M} \otimes_{\min} L^\infty(S, \mathcal{I}) \to \mathcal{M}$ the corresponding unital binormal CP map, i.e., $\Psi_{\mathcal{I}}$ is normal on $\mathcal{M}$ and $L^\infty(S, \mathcal{I})$ and satisfies $\Psi_{\mathcal{I}}(M \otimes [\chi_\Delta]) = \mathcal{I}(M, \Delta)$ for all $M \in \mathcal{M}$ and $\Delta \in \mathcal{F}$ [21, Proposition 3.3]. $\mathcal{I}$ is said to have the normal extension property (NEP) if there exists a unital normal CP map $\widetilde{\Psi_{\mathcal{I}}} : \mathcal{M}\overline{\otimes}L^\infty(S,\mathcal{I}) \to \mathcal{M}$ such that $\widetilde{\Psi_{\mathcal{I}}}|_{\mathcal{M}\otimes_{\min}L^\infty(S,\mathcal{I})} = \Psi_{\mathcal{I}}$. $\mathrm{CPInst}_{\mathrm{NE}}(\mathcal{M}, S)$ denotes the set of CP instruments for $(\mathcal{M}, S)$ with the NEP.

We then have the following theorem, a generalization of Theorem 19.

**Theorem 20** ([21, Theorems 3.4 and 3.6]) *For a CP instrument $\mathcal{I}$ for $(\mathcal{M}, S)$, the following conditions are equivalent:*
*(i) $\mathcal{I}$ has the NEP.*
*(ii) There exists CP instrument $\mathcal{I}$ for $(\mathbb{B}(\mathcal{H}), S)$ such that $\widetilde{\mathcal{I}}(M, \Delta) = \mathcal{I}(M, \Delta)$ for all $M \in \mathcal{M}$ and $\Delta \in \mathcal{F}$.*
*(iii) There exists a measuring process $\mathbb{M} = (\mathcal{K}, \sigma, E, U)$ for $(\mathcal{M}, S)$ such that $\mathcal{I}(M, \Delta) = \mathcal{I}_{\mathbb{M}}(M, \Delta)$ for all $M \in \mathcal{M}$ and $\Delta \in \mathcal{F}$.*

*The relation $\mathcal{I}(M, \Delta) = \mathcal{I}_{\mathbb{M}}(M, \Delta)$, where $M \in \mathcal{M}$ and $\Delta \in \mathcal{F}$, sets up a one-to-one correspondence between CP instruments $\mathcal{I}$ for $(\mathcal{M}, S)$ with the NEP and statistical equivalence classes of measuring processes $\mathbb{M} = (\mathcal{K}, \sigma, E, U)$ for $(\mathcal{M}, S)$.*

It is also shown that all CP instruments defined on a von Neumann algebra $\mathcal{M}$ have the NEP, i.e., $\mathrm{CPInst}(\mathcal{M}, S) = \mathrm{CPInst}_{\mathrm{NE}}(\mathcal{M}, S)$ if $\mathcal{M}$ is atomic [21, Theorem 4.1]. A von Neumann algebra $\mathcal{M}$ on a Hilbert space $\mathcal{H}$ is atomic if and only if there exists a normal conditional expectation $\mathcal{E} : \mathbb{B}(\mathcal{H}) \to \mathcal{M}$ [14, Chap. V, Sect. 2, Excercise 8].

In quantum mechanics, the complete positivity of instruments is physically justified in [22, 23] by considering a natural extendability, called *the trivial extendability*, of an instrument $\mathcal{I}$ on the system $\mathbf{S}$ to that $\mathcal{I}'$ on the composite system $\mathbf{S} + \mathbf{S}'$ containing the original one $\mathbf{S}$, where $\mathbf{S}'$ is an arbitrary system not interacting with $\mathbf{S}$ nor $\mathbf{A}(\mathbf{x})$. This justification of complete positivity is obtained as a part of an axiomatic characterization of physically realizable measurements [22, 23]. Then Theorem 19 enables us to regard the Davies-Lewis proposal restricted to CP instruments as a statement that is consistent with the standard formulation of quantum mechanics and hence acceptable for physicists.

#### 5.3.2.2 Repeatability and CP Instruments Without the NEP

First, we shall define the repeatability, the weak repeatability and the discreteness of instruments.

**Definition 25** ([21, *Definition* 5.3]) (1) An instrument $\mathcal{I}$ for $(\mathcal{M}, S)$ is said to be repeatable if $\mathcal{I}(\Delta_2)\mathcal{I}(\Delta_1) = \mathcal{I}(\Delta_2 \cap \Delta_1)$ for all $\Delta_1, \Delta_2 \in \mathcal{F}$.
(2) An instrument $\mathcal{I}$ for $(\mathcal{M}, S)$ is said to be weakly repeatable if $\mathcal{I}(\mathcal{I}(1, \Delta_2), \Delta_1) = \mathcal{I}(1, \Delta_2 \cap \Delta_1)$ for all $\Delta_1, \Delta_2 \in \mathcal{F}$.
(3) An instrument $\mathcal{I}$ for $(\mathcal{M}, S)$ is said to be discrete if there exist a countable subset $S_0$ of $S$ and a map $T : S_0 \to P(\mathcal{M}_*)$ such that

$$\mathcal{I}(\Delta)=\sum_{s\in\Delta}T(s) \tag{5.46}$$

for all $\Delta \in \mathcal{F}$.

It is easily seen that every repeatable instrument is weakly repeatable. Then we have the following proposition.

**Proposition 6** ([21, Proposition 5.9]) *Let $\mathcal{M}$ be a von Neumann algebra on a Hilbert space $\mathcal{H}$ and $(S, \mathcal{F})$ a measurable space. Every discrete CP instrument $\mathcal{I}$ for $(\mathcal{M}, S)$ has the NEP.*

This fact is implicitly used later. By contrast, we have the structure theorem for the weakly repeatable CP instruments.

**Theorem 21** ([21, Theorem 5.10]) *Let $\mathcal{M}$ be a von Neumann algebra on a Hilbert space $\mathcal{H}$ and $(S, \mathcal{F})$ a standard Borel space. A weakly repeatable CP instrument $\mathcal{I}$ for $(\mathcal{M}, S)$ is discrete if and only if it has the NEP.*

To show this theorem, we need to develop the theory of posterior states for instruments. Here we omit it and refer the reader to [21, 33] for its details.

In the case where $\mathcal{M}$ is non-atomic, there exist CP instruments for $(\mathcal{M}, S)$ without the NEP. The following two CP instruments are such examples.

*Example 8* ([33, pp. 292–293], [21, Example 5.1]) Let $m$ be Lebesgue measure on $[0, 1]$. A CP instrument $\mathcal{I}_m$ for $(L^\infty([0, 1], m), [0, 1])$ is defined by $\mathcal{I}_m(f, \Delta) = [\chi_\Delta]f$ for all $\Delta \in \mathcal{B}([0, 1])$ and $f \in L^\infty([0, 1], m)$.

We say that a von Neumann algebra $\mathcal{M}$ is approximately finite-dimensional (AFD) if there is an increasing net $\{\mathcal{M}_\alpha\}_{\alpha\in A}$ of finite-dimensional von Neumann subalgebras of $\mathcal{M}$ such that

$$\mathcal{M}=\overline{\bigcup_{\alpha\in A}\mathcal{M}_\alpha}^{uw}. \tag{5.47}$$

*Example 9* ([21, Example 5.2]) Let $\mathcal{M}$ be an AFD von Neumann algebra of type $\mathrm{II}_1$ on a separable Hilbert space $\mathcal{H}$. Let $A = \int_{\mathbb{R}} a\, dE^A(a)$ be a self-adjoint operator with continuous spectrum affiliated with $\mathcal{M}$ and $\mathcal{E}$ a (normal) conditional expectation of $\mathcal{M}$ onto $\{A\}' \cap \mathcal{M}$ (the existence of $\mathcal{E}$ was first found by [34, Theorem 1]), where $\{A\}' = \{E^A(\Delta) \mid \Delta \in \mathcal{B}(\mathbb{R})\}'$. A CP instrument $\mathcal{I}_A$ for $(\mathcal{M}, \mathbb{R})$ is defined by

$$\mathcal{I}_A(M, \Delta) = \mathcal{E}(M)E^A(\Delta) \tag{5.48}$$

for all $M \in \mathcal{M}$ and $\Delta \in \mathcal{B}(\mathbb{R})$.

By Theorem 21, the weak repeatability and the non-discreteness of $\mathcal{I}_m$ and $\mathcal{I}_A$ imply the non-existence of measuring processes which define them. These examples are very important for the dilation theory of CP maps since they revealed the existence of families of CP instruments which do not admit unitary dilations.

#### 5.3.2.3 Measurement Theory in the Heisenberg Picture

To discuss processes of quantum measurement in the Heisenberg picture, we introduce the concept of system of measurement correlations. It is a multivariate version of instrument and is defined as a family of multilinear maps satisfying "positive-definiteness", "$\sigma$-additivity" and other conditions. The theory of quantum stochastic processes by Accardi, Frigerio and Lewis [35] then has an affinity for systems of measurement correlations, which describes quantum dynamical changes of the system by families of $^*$-homomorphisms indexed by time parameters. We use the theory of quantum stochastic processes and a noncommutative Kolmogorov's theorem to show the existence of measuring processes which defines a given system of measurement correlations defined on $\mathbb{B}(\mathcal{H})$. Our approach considers quantum systems with infinite degrees of freedom described by general von Neumann algebras in order to comprehensively understand physics of quantum measurement in the Heisenberg picture. The author thinks that the understanding of the (usual) Heisenberg picture has not been deepened in contrast to the Schrödinger picture. It should be stressed that the circumstances are never restricted to quantum measurement theory.

To begin with, we shall introduce the notations below.

**Notation 1** *Let $\mathcal{T}^{(1)}$ be a set. We define a set $\mathcal{T}$ by $\mathcal{T} = \cup_{j=1}^{\infty}(\mathcal{T}^{(1)})^j$.*
*(i) For each $T \in \mathcal{T}$, $|T|$ denotes the natural number $n$ such that $T \in (\mathcal{T}^{(1)})^n$.*
*(ii) For each $T = (t_1, t_2, \ldots, t_{n-1}, t_n) \in \mathcal{T}$, we define $T^\# \in \mathcal{T}$ by $T^\# = (t_n, t_{n-1}, \ldots, t_2, t_1)$.*
*(iii) For any $T = (t_{1,1}, \ldots, t_{1,m}), T_2 = (t_{2,1}, \ldots, t_{2,n}) \in \mathcal{T}$, the product $T_1 \times T_2$ is defined by*

$$T_1 \times T_2 = (t_{1,1}, \ldots, t_{1,m}, t_{2,1}, \ldots, t_{2,n}). \tag{5.49}$$

*Since it holds that $T_1 \times (T_2 \times T_3) = (T_1 \times T_2) \times T_3$, $T_1 \times (T_2 \times T_3)$ is written as $T_1 \times T_2 \times T_3$.*
*(iv) For any $n \in \mathbb{N}$ and $\overrightarrow{M} = (M_1, M_2, \ldots, M_{n-1}, M_n) \in \mathcal{M}^n$, we define $\overrightarrow{M}^\# \in \mathcal{M}^n$ by*

$$\overrightarrow{M}^\# = (M_n^*, M_{n-1}^*, \ldots, M_2^*, M_1^*). \tag{5.50}$$

*(v) For any $m, n \in \mathbb{N}$, $\overrightarrow{M}_1 = (M_{1,1}, \ldots, M_{1,m}) \in \mathcal{M}^m$ and $\overrightarrow{M}_2 = (M_{2,1}, \ldots, M_{2,n}) \in \mathcal{M}^n$, the product $\overrightarrow{M}_1 \times \overrightarrow{M}_2 \in \mathcal{M}^{m+n}$ is defined by*

$$\overrightarrow{M}_1 \times \overrightarrow{M}_2 = (M_{1,1}, \ldots, M_{1,m}, M_{2,1}, \ldots, M_{2,n}). \tag{5.51}$$

*Since it holds that $\overrightarrow{M}_1 \times (\overrightarrow{M}_2 \times \overrightarrow{M}_3) = (\overrightarrow{M}_1 \times \overrightarrow{M}_2) \times \overrightarrow{M}_3$, $\overrightarrow{M}_1 \times (\overrightarrow{M}_2 \times \overrightarrow{M}_3)$ is written as $\overrightarrow{M}_1 \times \overrightarrow{M}_2 \times \overrightarrow{M}_3$.*
*(vi) For every family $\{\Pi_t\}_{t\in\mathcal{T}^{(1)}}$ of representations of $\mathcal{M}$ on a Hilbert space $\mathcal{L}$, we adopt the notation*

$$\Pi_T(\overrightarrow{M}) = \Pi_{t_1}(M_1)\cdots\Pi_{t_{|T|}}(M_{|T|}) \tag{5.52}$$

*for all* $T = (t_1, \ldots, t_{|T|}) \in \mathcal{T}$ *and* $\overrightarrow{M} = (M_1, \ldots, M_{|T|}) \in \mathcal{M}^{|T|}$.

Let $(S, \mathcal{F})$ be a measurable space. We define a set $\mathcal{T}_S$ by

$$\mathcal{T}_S = \cup_{j=1}^{\infty} (\mathcal{T}_S^{(1)})^j, \tag{5.53}$$

$$\mathcal{T}_S^{(1)} = \{in\} \cup \mathcal{F}, \tag{5.54}$$

where $in$ is a symbol.

A system of measurement correlations is defined as an "operator-valued" version of projective system of multikernels analyzed in the previous investigations [35, 36], which satisfies the positive-definiteness, a representation theoretically fruitful property. We treat only the case that systems of measurement correlations do not have explicit time-dependence for simplicity herein.

**Definition 26** (*System of measurement correlations* [24, Definition 8]) A family $\{W_T\}_{T\in\mathcal{T}}$ of maps $W_T : \mathcal{M}^{|T|} = \overbrace{\mathcal{M} \times \cdots \times \mathcal{M}}^{|T|} \to \mathcal{M}$ is called for $(\mathcal{M}, S)$ if it satisfies $\mathcal{T}^{(1)} = \mathcal{T}_S^{(1)}$ and the following six conditions:
(MC1) For any $T \in \mathcal{T}$, $W_T(M_1, \ldots, M_{|T|})$ is separately linear and ultraweakly continuous in each variable $M_1, \ldots, M_{|T|} \in \mathcal{M}$.
(MC2) For any $n \in \mathbb{N}$, $(T_1, \overrightarrow{M}_1), \ldots, (T_n, \overrightarrow{M}_n) \in \cup_{T\in\mathcal{T}}(\{T\} \times \mathcal{M}^{|T|})$, and $\xi_1, \ldots, \xi_n \in \mathcal{H}$,

$$\sum_{i,j=1}^{n} \langle \xi_i | W_{T_i^{\#}\times T_j}(\overrightarrow{M}_i^{\#} \times \overrightarrow{M}_j)\xi_j\rangle \geq 0. \tag{5.55}$$

(MC3) For any $T = (t_1, \ldots, t_{|T|}) \in \mathcal{T}$, $\overrightarrow{M} = (M_1, \ldots, M_{|T|}) \in \mathcal{M}^{|T|}$ and $M \in \mathcal{M}$,

$$M W_T(\overrightarrow{M}) = W_{(in)\times T}((M) \times \overrightarrow{M}), \tag{5.56}$$

$$W_T(\overrightarrow{M})M = W_{T\times(in)}(\overrightarrow{M} \times (M)). \tag{5.57}$$

(MC4) Let $T = (t_1, \ldots, t_{|T|}) \in \mathcal{T}$. If $t_k = t_{k+1} = in$ or $t_k, t_{k+1} \in \mathcal{F}$ for some $1 \leq k \leq |T| - 1$,

$$W_T(M_1, \ldots, M_k, M_{k+1}, \ldots, M_{|T|}) = W_{T'}(M_1, \ldots, M_k M_{k+1}, \ldots, M_{|T|}) \tag{5.58}$$

for all $(M_1, \ldots, M_{|T|}) \in \mathcal{M}^{|T|}$, where $T' = (t_1, \ldots, t_{k-1}, t_k \cap t_{k+1}, t_{k+2} \ldots, t_{|T|})$ and

$$t_k \cap t_{k+1} = \begin{cases} in, & (\text{if } t_k = t_{k+1} = in) \\ t_k \cap t_{k+1}, & (\text{if } t_k, t_{k+1} \in \mathcal{F}). \end{cases}$$

(MC5) For any $T = (t_1, \ldots, t_{|T|}) \in \mathcal{T}$ with $t_k = in$ or $S$, and $(M_1, \ldots, M_{|T|}) \in \mathcal{M}^{|T|}$ with $M_k = 1$,

$$W_T(M_1,\ldots,M_{k-1},1,M_{k+1},\ldots,M_{|T|}) = W_{\hat{k}T}(M_1,\ldots,M_{k-1},M_{k+1},\ldots,M_{|T|}), \tag{5.59}$$

where $\hat{k}T = (t_1,\ldots,t_{k-1},t_{k+1},t_{k+2}\ldots,t_{|T|})$. In addition, $W_{in}(1) = W_S(1) = 1$.
(MC6) For any $n \in \mathbb{N}$, $1 \leq k \leq n$, $t_1,\ldots,t_{k-1},t_{k+1},\ldots,t_n \in \mathcal{T}^{(1)}$, mutually disjoint sequence $\{t_{k,j}\}_j \subset \mathcal{F}$, $\overrightarrow{M} \in \mathcal{M}^n$ and $\rho \in \mathcal{M}_*$,

$$\langle \rho, W_{(t_1,\ldots,t_{k-1},\cup_j t_{k,j},t_{k+1},\ldots,t_n)}(\overrightarrow{M})\rangle = \sum_j \langle \rho, W_{(t_1,\ldots,t_{k-1},t_{k,j},t_{k+1},\ldots,t_n)}(\overrightarrow{M})\rangle. \tag{5.60}$$

$in$ and $\Delta \in \mathcal{F}$ are subscripts that specify the time before the measurement and the time after the measurement, respectively. In each $W_T(\overrightarrow{M})$, where $T \in \mathcal{T}$ and $\overrightarrow{M} \in \mathcal{M}^{|T|}$, components of $\overrightarrow{M}$ indexed by $in$ and those of $\overrightarrow{M}$ indexed by $\Delta \in \mathcal{F}$, describe observables before the measurement and those after the measurement, respectively. In particular, the latter precisely describe observables of the system after the measurement in the situation that values of the output variable of the measuring apparatus are restricted to $\Delta \in \mathcal{F}$. This understanding of systems of measurement correlations is supported by Theorem 22.

We shall newly define measuring processes consistent with systems of measurement correlations.

**Definition 27** ([24, *Definition* 9]) A measuring process $\mathbb{M}$ for $(\mathcal{M}, S)$ is a 4-tuple $\mathbb{M} = (\mathcal{K}, \sigma, E, U)$ which consists of a Hilbert space $\mathcal{K}$, a normal state $\sigma$ on $\mathbb{B}(\mathcal{K})$, a spectral measure $E : \mathcal{F} \to \mathbb{B}(\mathcal{K})$, and a unitary operator $U$ on $\mathcal{H} \otimes \mathcal{K}$ and defines a system of measurement correlations $\{W_T^{\mathbb{M}}\}_{T\in\mathcal{T}_S}$ for $(\mathcal{M}, S)$ as follows: We define a representation $\pi_{in}$ of $\mathcal{M}$ and a family $\{\pi_\Delta\}_{\Delta\in\mathcal{F}}$ of those of $\mathcal{M}$ on $\mathcal{H} \otimes \mathcal{K}$ by

$$\pi_{in}(M) = M \otimes 1_{\mathcal{K}}, \quad \pi_\Delta(M) = U^*(M \otimes E(\Delta))U \tag{5.61}$$

for all $M \in \mathcal{M}$ and $\Delta \in \mathcal{F}$, respectively. For each $T \in \mathcal{T}_S$, $W_T^{\mathbb{M}} : \mathcal{M}^{|T|} \to \mathcal{M}$ is defined by

$$W_T^{\mathbb{M}}(\overrightarrow{M}) = (\mathrm{id} \otimes \sigma)[\pi_T(\overrightarrow{M})] \tag{5.62}$$

for all $\overrightarrow{M} \in \mathcal{M}^{|T|}$.

We see that two definitions of measuring processes for $(\mathbb{B}(\mathcal{H}), S)$ are equivalent.

**Definition 28** We say that a CP instrument $\mathcal{I}$ for $(\mathcal{M}, S)$ is realized by a measuring process $\mathbb{M}$ for $(\mathcal{M}, S)$ in the sense of Definition 27, or $\mathbb{M}$ realizes $\mathcal{I}$ if $\mathcal{I} = \mathcal{I}_{\mathbb{M}}$. $\mathrm{CPInst}_{\mathrm{RE}}(\mathcal{M}, S)$ denotes the set of CP instruments for $(\mathcal{M}, S)$ realized by measuring processes for $(\mathcal{M}, S)$ in the sense of Definition 27.

Surprisingly, we have the following proposition.

**Proposition 7** *Let $\mathcal{M}$ be a von Neumann algebra on a Hilbert space $\mathcal{H}$ and $(S, \mathcal{F})$ a measurable space. Then we have*

$$\mathrm{CPInst}_{\mathrm{RE}}(\mathcal{M}, S) = \mathrm{CPInst}_{\mathrm{NE}}(\mathcal{M}, S). \tag{5.63}$$

**Definition 29** ([24, *Definition 10*]) Let $n \in \mathbb{N}$. Two measuring processes $\mathbb{M}_1$ and $\mathbb{M}_2$ for $(\mathcal{M}, S)$ are said to be $n$-equivalent if $W_T^{\mathbb{M}_1} = W_T^{\mathbb{M}_2}$ for all $T \in \mathcal{T}$ such that $|T| \leq n$. Two measuring processes $\mathbb{M}_1$ and $\mathbb{M}_2$ for $(\mathcal{M}, S)$ are said to be completely equivalent if they are $n$-equivalent for all $n \in \mathbb{N}$.

For every system of measurement correlations for $(\mathbb{B}(\mathcal{H}), S)$, the following realization theorem by measuring processes holds.

**Theorem 22** ([24, Theorem 9]) *Let $\mathcal{H}$ be a Hilbert space and $(S, \mathcal{F})$ a measurable space. Then there is a one-to-one correspondence between complete equivalence classes of measuring processes $\mathbb{M} = (\mathcal{K}, \sigma, E, U)$ for $(\mathbb{B}(\mathcal{H}), S)$ and systems $\{W_T\}_{T\in\mathcal{T}}$ of measurement correlations for $(\mathbb{B}(\mathcal{H}), S)$, which is given by the relation*

$$W_T(\overrightarrow{M}) = W_T^{\mathbb{M}}(\overrightarrow{M}) \tag{5.64}$$

*for all $T \in \mathcal{T}$ and $\overrightarrow{M} \in \mathcal{M}^{|T|}$.*

It is obvious that every system of measurement correlations defines a CP instrument. The "converse" is also true:

**Theorem 23** ([24, Theorem 10]) *Let $\mathcal{M}$ be a von Neumann algebra on a Hilbert space $\mathcal{H}$ and $(S, \mathcal{F})$ a measurable space. For every CP instrument $\mathcal{I}$ for $(\mathcal{M}, S)$, there exists a system of measurement correlations $\{W_T\}_{T\in\mathcal{T}_S}$ for $(\mathcal{M}, S)$ such that*

$$\mathcal{I}(M, \Delta) = \mathcal{I}_W(M, \Delta) \tag{5.65}$$

*for all $M \in \mathcal{M}$ and $\Delta \in \mathcal{F}$.*

In the case where $\mathcal{M}$ is an atomic von Neumann algebra on a Hilbert space $\mathcal{H}$, we have another construction of a system of measurement correlations which defines a given CP instrument (See [24]).

#### 5.3.2.4 Approximate Realizability of CP Instruments by Measuring Processes

Here we discuss the approximate realizability of CP instruments by measuring processes, equivalently, by CP instruments with the NEP. This is motivated by the existence of CP instruments without the NEP.

**Definition 30** (*Approximately normal extension property*) A CP instrument $\mathcal{I}$ for $(\mathcal{M}, S)$ is said to have the approximately normal extension property (ANEP) if there is a net $\{\mathcal{I}_\alpha\}$ of CP instruments with the NEP such that $\mathcal{I}_\alpha(M, \Delta)$ is ultraweakly converges to $\mathcal{I}(M, \Delta)$ for all $M \in \mathcal{M}$ and $\Delta \in \mathcal{F}$. $\mathrm{CPInst}_{\mathrm{AN}}(\mathcal{M}, S)$ denotes the set of CP instruments for $(\mathcal{M}, S)$ with the ANEP.

The following theorem holds for general $\sigma$-finite von Neumann algebras without assuming any other conditions.

**Theorem 24** *Let $\mathcal{M}$ be a von Neumann algebra on a Hilbert space $\mathcal{H}$ and $(S, \mathcal{F})$ a measurable space. For every CP instrument $\mathcal{I}$ for $(\mathcal{M}, S)$, $n \in \mathbb{N}$, $\rho_1, \ldots, \rho_n \in \mathcal{S}_n(\mathcal{M})$, $M_1, \ldots, M_n \in \mathcal{M}$ and $\Delta_1, \ldots, \Delta_n \in \mathcal{F}$, there exists a measuring process $\mathbb{M} = (\mathcal{K}, \sigma, E, U)$ for $(\mathcal{M}, S)$ such that*

$$\langle \rho_j, \mathcal{I}(M_j, \Delta_j)\rangle = \langle \rho_j, \mathcal{I}_{\mathbb{M}}(M_j, \Delta_j)\rangle \tag{5.66}$$

*for all $j = 1, \ldots, n$.*

**Corollary 5** *Let $\mathcal{M}$ be a von Neumann algebra on a Hilbert space $\mathcal{H}$ and $(S, \mathcal{F})$ a measurable space. Then we have*

$$\mathrm{CPInst}_{\mathrm{AN}}(\mathcal{M}, S) = \mathrm{CPInst}(\mathcal{M}, S). \tag{5.67}$$

**Definition 31** A CP instrument $\mathcal{I}$ for $(\mathcal{M}, S)$ is said to be approximately realized by a net of measuring processes $\{\mathbb{M}_\alpha\}_{\alpha\in A}$ for $(\mathcal{M}, S)$, or $\{\mathbb{M}_\alpha\}_{\alpha\in A}$ approximately realizes $\mathcal{I}$ if, for every $\varepsilon > 0$, $n \in \mathbb{N}$, $\rho_1, \ldots, \rho_n \in \mathcal{S}_n(\mathcal{M})$, $\Delta_1, \ldots, \Delta_n \in \mathcal{F}$ and $M_1, \ldots, M_n \in \mathcal{M}$, there is $\alpha \in A$ such that $|\langle \rho_i, \mathcal{I}(M_i, \Delta_i)\rangle - \langle \rho_i, \mathcal{I}_{\mathbb{M}_\alpha}(M_i, \Delta_i)\rangle| < \varepsilon$ for all $i = 1, \ldots, n$. $\mathrm{CPInst}_{\mathrm{AR}}(\mathcal{M}, S)$ denotes the set of CP instruments for $(\mathcal{M}, S)$ approximately realized by nets of measuring processes for $(\mathcal{M}, S)$.

By Proposition 7 and Corollary 5, we have

$$\mathrm{CPInst}_{\mathrm{AR}}(\mathcal{M}, S) = \mathrm{CPInst}_{\mathrm{AN}}(\mathcal{M}, S) = \mathrm{CPInst}(\mathcal{M}, S). \tag{5.68}$$

Next, we extend the (unfinished) program, advocated and developed by many researchers [37–40], which states that physical processes should be described by (inner) CP maps usually called operations [37, 41] or effects [39].

**Definition 32** (*Approximate innerness* [42, 43]) Let $\mathcal{M}$ be a von Neumann algebra on a Hilbert space $\mathcal{H}$.
(1) A positive linear map $\Psi$ of $\mathcal{M}$ is said to be finitely inner if there is a finite sequence $\{V_j\}_{j=1,\ldots,m}$ of $\mathcal{M}$ such that $\Psi(M) = \sum_{j=1}^{m} V_j^* M V_j$ for all $M \in \mathcal{M}$
(2) A positive linear map $\Psi$ of $\mathcal{M}$ is said to be inner if there is a sequence $\{V_j\}_{j\in\mathbb{N}}$ of $\mathcal{M}$ such that $\Psi(M) = \sum_{j=1}^{\infty} V_j^* M V_j$ for all $M \in \mathcal{M}$, where the convergence is ultraweak.
(3) A positive linear map $\Psi$ of $\mathcal{M}$ is said to be approximately inner if it is the pointwise ultraweak limit of a net $\{\Psi_\alpha\}_{\alpha\in A}$ of finitely inner positive linear maps such that $\Psi_\alpha(1) \leq \Psi(1)$ for all $\alpha \in A$.

Anantharaman–Delaroche and Havet [42] called the finite innerness (and the approximate innerness, respectively) of CP maps the factorization through the identity map $\mathrm{id}_{\mathcal{M}} : \mathcal{M} \to \mathcal{M}$ (and the approximate factorization through $\mathrm{id}_{\mathcal{M}}$, respectively). We refer the reader to [42–45] for more detailed discussions. Every finitely inner positive linear map $\Psi$ of $\mathcal{M}$ is of course inner. In addition, every inner positive linear map $\Psi(M) = \sum_{j=1}^{\infty} V_j^* M V_j$, $M \in \mathcal{M}$, is approximately inner since it is the ultraweak limit of a sequence $\{\Psi_j\}$ of finitely inner positive maps $\Psi_j(M) = \sum_{k=1}^{j} V_k^* M V_k$, $M \in \mathcal{M}$, such that $\Psi_j(1) = \sum_{k=1}^{j} V_k^* V_k \leq \Psi(1)$ for all $j \in \mathbb{N}$. Every approximately inner positive linear map $\Psi$ of $\mathcal{M}$ is always completely positive.

The following proposition states that the program of interest is affirmatively valid.

**Proposition 8** (Anantharaman–Delaroche and Havet [42, Lemma 2.2, Remarks 5.4]) *Let $\mathcal{M}$ be an injective factor on on a Hilbert space $\mathcal{H}$. Every CP map $\Psi$ of $\mathcal{M}$ is approximately inner, i.e., it is the pointwise ultraweak limit of a net of CP maps $\{\Psi_\theta\}_{\theta\in\Theta}$ of the form $\Psi_\theta(M) = \sum_{j=1}^{n_\theta} V_{\theta,j}^* M V_{\theta,j}$, $M \in \mathcal{M}$, with $n_\theta \in \mathbb{N}$, $V_{\theta,1}, \ldots, V_{\theta,n_\theta} \in \mathcal{M}$ such that $\Psi_\theta(1) = \sum_{j=1}^{n_\theta} V_{\theta,j}^* V_{\theta,j} \leq \Psi(1)$ for all $\theta \in \Theta$.*

It is shown in [42] that counterexamples exist for injective non-factors. We will see that Theorem 26 is exactly a goal of the program.

**Definition 33** An instrument $\mathcal{I}$ for $(\mathcal{M}, S)$ is said to be finitely inner [inner, or approximately inner, respectively] if $\mathcal{I}(\cdot, \Delta)$ is finitely inner [inner, or approximately inner, respectively] for every $\Delta \in \mathcal{F}$. $\mathrm{CPInst}_{\mathrm{FI}}(\mathcal{M}, S)$ [$\mathrm{CPInst}_{\mathrm{IN}}(\mathcal{M}, S)$, or $\mathrm{CPInst}_{\mathrm{AI}}(\mathcal{M}, S)$, respectively] denotes the set of finitely inner [inner, or approximately inner, respectively] CP instruments for $(\mathcal{M}, S)$.

The following relation holds.

$$\mathrm{CPInst}_{\mathrm{FI}}(\mathcal{M}, S) \subset \mathrm{CPInst}_{\mathrm{IN}}(\mathcal{M}, S) \subset \mathrm{CPInst}_{\mathrm{AI}}(\mathcal{M}, S). \tag{5.69}$$

**Definition 34** (*Inner measuring process*) We say that a measuring process $\mathbb{M} = (\mathcal{K}, \sigma, E, U)$ for $(\mathcal{M}, S)$ is inner if $U$ is contained in $\mathcal{M}\overline{\otimes}\mathbb{B}(\mathcal{K})$.

We then have the following theorem.

**Theorem 25** ([24, Theorem 11]) *Let $\mathcal{M}$ be a von Neumann algebra on a Hilbert space $\mathcal{H}$ and $(S, \mathcal{F})$ a measurable space. For every approximately inner (hence CP) instrument $\mathcal{I}$ for $(\mathcal{M}, S)$, $\varepsilon > 0$, $n \in \mathbb{N}$, $\rho_1, \ldots, \rho_n \in \mathcal{S}_n(\mathcal{M})$, $M_1, \ldots, M_n \in \mathcal{M}$ and $\Delta_1, \ldots, \Delta_n \in \mathcal{F}$, there exists an inner measuring process $\mathbb{M} = (\mathcal{K}, \sigma, E, U)$ for $(\mathcal{M}, S)$ in the sense of Definition 27 such that*

$$|\langle \rho_j, \mathcal{I}(M_j, \Delta_j)\rangle - \langle \rho_j, \mathcal{I}_{\mathbb{M}}(M_j, \Delta_j)\rangle| < \varepsilon \tag{5.70}$$

*for all $j = 1, \ldots, n$.*

Only for injective factors, the following theorem is established as a corollary of the above theorem.

**Theorem 26** ([24, Theorem 12]) *Let $\mathcal{M}$ be an injective factor on a Hilbert space $\mathcal{H}$ and $(S, \mathcal{F})$ a measurable space. For every CP instrument $\mathcal{I}$ for $(\mathcal{M}, S)$, $\varepsilon > 0, n \in \mathbb{N}$, $\rho_1, \ldots, \rho_n \in \mathcal{S}_n(\mathcal{M})$, $M_1, \ldots, M_n \in \mathcal{M}$ and $\Delta_1, \ldots, \Delta_n \in \mathcal{F}$, there exists an inner measuring process $\mathbb{M} = (\mathcal{K}, \sigma, E, U)$ for $(\mathcal{M}, S)$ in the sense of Definition 27 such that*

$$|\langle \rho_j, \mathcal{I}(M_j, \Delta_j)\rangle - \langle \rho_j, \mathcal{I}_{\mathbb{M}}(M_j, \Delta_j)\rangle| < \varepsilon \tag{5.71}$$

*for all $j = 1, \ldots, n$.*

**Corollary 6** ([24, Corollary 5]) *Let $\mathcal{M}$ be an injective factor on a Hilbert space $\mathcal{H}$ and $(S, \mathcal{F})$ a measurable space. Then we have*

$$\mathrm{CPInst}_{\mathrm{AI}}(\mathcal{M}, S) = \mathrm{CPInst}(\mathcal{M}, S). \tag{5.72}$$

In the context of instrument, it is natural to consider that we regard Theorem 26 as an accomplishment of the program of interest for injective factors.

**Definition 35** ([24, *Definition 14*]) We say that a measuring process $\mathbb{M} = (\mathcal{K}, \sigma, E, U)$ for $(\mathcal{M}, S)$ is faithful if there exists a normal faithful representation $\widetilde{E} : L^\infty(S, \mathcal{I}_{\mathbb{M}}) \to \mathbb{B}(\mathcal{K})$ such that $\widetilde{E}([\chi_\Delta]) = E(\Delta)$ for all $\Delta \in \mathcal{F}$.

This definition is the same as [21, Definition 3.4] except that the definition of measuring process for $(\mathcal{M}, S)$ is different.

**Theorem 27** ([24, Theorem 13]) *Let $\mathcal{M}$ be an injective von Neumann algebra on a Hilbert space $\mathcal{H}$ and $(S, \mathcal{F})$ a measurable space. For every CP instrument $\mathcal{I}$ for $(\mathcal{M}, S)$, $\varepsilon > 0$, $n \in \mathbb{N}$, $\rho_1, \ldots, \rho_n \in \mathcal{S}_n(\mathcal{M})$, $M_1, \ldots, M_n \in \mathcal{M}$ and $\Delta_1, \ldots, \Delta_n \in \mathcal{F}$, there exists a faithful measuring process $\mathbb{M} = (\mathcal{K}, \sigma, E, U)$ for $(\mathcal{M}, S)$ in the sense of Definition 27 such that*

$$|\langle \rho_j, \mathcal{I}(M_j, \Delta_j)\rangle - \langle \rho_j, \mathcal{I}_{\mathbb{M}}(M_j, \Delta_j)\rangle| < \varepsilon \tag{5.73}$$

*for all $j = 1, \ldots, n$, and that $\mathcal{I}(1, \Delta) = \mathcal{I}_{\mathbb{M}}(1, \Delta)$ for all $\Delta \in \mathcal{F}$.*

By the proof of Theorem 27 and Proposition 7, we have Corollary 5 and the following.

**Corollary 7** *Let $\mathcal{M}$ be an atomic von Neumann algebra on a Hilbert space $\mathcal{H}$ and $(S, \mathcal{F})$ a measurable space. Then we have*

$$\mathrm{CPInst}_{\mathrm{RE}}(\mathcal{M}, S) = \mathrm{CPInst}(\mathcal{M}, S). \tag{5.74}$$

The solution of the following question will be the future task.

*Question 1* Let $\mathcal{M}$ be an injective factor on a Hilbert space $\mathcal{H}$. For every CP instrument $\mathcal{I}$ for $(\mathcal{M}, S)$, does there exist a net $\{\mathbb{M}_\alpha\}_{\alpha \in A}$ of inner faithful measuring processes for $(\mathcal{M}, S)$ such that $\mathcal{I}_{\mathbb{M}_\alpha} \to^{uw} \mathcal{I}$ and $\mathcal{I}_{\mathbb{M}_\alpha}(1, \Delta) = \mathcal{I}(1, \Delta)$ for all $\Delta \in \mathcal{F}$?

In physically relevant cases, it is known that every von Neumann algebra $\mathcal{M}$ describing the observable algebra of a quantum system acts on a separable Hilbert space and is AFD. For example, it is shown in [46] that von Neumann algebras of local observables in quantum field theory are AFD and acts on a separable Hilbert space under natural postulates, e.g., the Wightman axioms, the nuclearity condition and the asymptotic scale invariance. For every von Neumann algebra $\mathcal{M}$ on a separable Hilbert space (or with separable dual, equivalently), $\mathcal{M}$ is AFD if and only if it is injective, furthermore, if and only if it is amenable [9, 15]. Hence the assumption of the injectivity for von Neumann algebras is very natural.

**Acknowledgements** The author thanks Prof. Ohtsu for promoting dressed photon, a very attractive research subject, and encouraging him warmly. He also thanks to other members of Research Origin for Dressed Photon.

This work is supported by the JSPS KAKENHI, No. 26247016, No. 16K17641, No. 17H01277, by Research Origin for Dressed Photon, by Grant for Basic Science Research Projects from The Sumitomo Foundation and by JST CREST Grant Number JPMJCR17N2, Japan.

## References

1. R.F. Streater, A.S. Wightman, *PCT, Spin and Statistics, and All That* (Princeton University press, Princeton, 2000)
2. M. Ohtsu, *Dressed Photons* (Springer, Berlin, 2016)
3. I. Ojima, A unified scheme for generalized sectors based on selection criteria order parameters of symmetries and of thermality and physical meanings of adjunctions. Open Sys. Inform. Dyn. **10**, 235–279 (2003)
4. I. Ojima, Temperature as order parameter of broken scale invariance. Publ. RIMS **40**, 731–756 (2004)
5. I. Ojima, Micro-macro duality in quantum physics, in *Proceedings of the International Conference on Stochastic Analysis, Classical and Quantum* (World Scientific, 2005), pp. 143–161. arXiv:math-ph/0502038
6. I. Ojima, Meaning of non-extensive entropies in micro-macro duality. J. Phys. **201**, 012017 (2010)
7. O. Bratteli, D.W. Robinson, *Operator Algebras and Quantum Statistical Mechanics*, vol. 1 (2nd printing of 2nd edn.) (Springer, Berlin, 2002)
8. O. Bratteli, D.W. Robinson, *Operator Algebras and Quantum Statistical Mechanics*, vol. 2 (2nd printing of 2nd edn.) (Springer, Berlin, 2002)
9. A. Connes, *Noncommutative Geometry* (Academic, San Diego, 1994)
10. J. Dixmier, *C*-Algebras* (North-Holland, Amsterdam, 1977)
11. J. Dixmier, *Von Neumann Algebras* (North-Holland, Amsterdam, 1981)
12. V. Paulsen, *Completely Bounded Maps and Operator Algebras* (Cambridge University press, Cambridge, 2002)
13. S. Sakai, *C*-Algebras and W*-Algebras* (Springer, Berlin, 1971)
14. M. Takesaki, *Theory of Operator Algebras I* (Springer, Berlin, 1979)
15. M. Takesaki, *Theory of Operator Algebras III* (Springer, Berlin, 2002)
16. G. Emch, *Algebraic Methods in Statistical Mechanics and Quantum Field Theory* (Wiley-Interscience, New York, 1972)
17. J. von Neumann, *Mathematische Grundlagen der Quantenmechanik* (Springer, Berlin, 1932); *Mathematical Foundations of Quantum Mechanics* (Princeton University Press, Princeton, 1955)

18. A. Hora, N. Obata, *Quantum Probability and Spectral Analysis of Graphs* (Springer, Berlin, 2007)
19. I. Ojima, K. Okamura, H. Saigo, Derivation of born rule from algebraic and statistical axioms. Open Sys. Inform. Dyn. **21**(3), 1450005 (2014). https://doi.org/10.1142/S123016121450005X
20. I. Ojima, K. Okamura, *Mathematics and Physics of Infinite Quantum Systems* (in Japanese), (Saiensu-sha, 2013)
21. K. Okamura, M. Ozawa, Measurement theory in local quantum physics. J. Math. Phys. **57**, 015209 (2016)
22. M. Ozawa, Uncertainty relations for noise and disturbance in generalized quantum measurements. Ann. Phys. (N.Y.) **331**, 350–416 (2004)
23. M. Ozawa, Mathematical foundations of quantum information: measurement and foundations. Sugaku Expo. **27**, 195–221 (2014)
24. K. Okamura, Measuring processes and the Heisenberg picture, arXiv:1511.09228 (2018)
25. M. Nakamura, H. Umegaki, On von Neumann's theory of measurements in quantum statistics. Math. Jpn. **7**, 151–157 (1962)
26. W. Arveson, Analyticity in operator algebras. Am. J. Math. **89**, 578–642 (1967)
27. E.B. Davies, J.T. Lewis, An operational approach to quantum probability. Commun. Math. Phys. **17**, 239–260 (1970)
28. M. Ozawa, Heisenberg's original derivation of the uncertainty principle and its universally valid reformulations. Curr. Sci. **10**, 2006–2016 (2015)
29. M. Ozawa, Conditional expectation and repeated measurements of continuous quantum observables, in *Probability Theory and Mathematical Statistics*, ed. by K. Ito and J.V. Prohorov. Lecture Notes Mathematics, vol. 1021 (Springer, Berlin, 1983), pp. 518–525
30. M. Ozawa, Quantum measuring processes of continuous observables. J. Math. Phys. **25**, 79–87 (1984)
31. M. Ozawa, An operational approach to quantum state reduction. Ann. Phys. (N.Y.) **259**, 121–137 (1997)
32. M. Ozawa, Canonical approximate quantum measurements. J. Math. Phys. **34**, 5596–5624 (1993)
33. M. Ozawa, Conditional probability and a posteriori states in quantum mechanics. Publ. Res. Inst. Math. Sci. **21**, 279–295 (1985)
34. H. Umegaki, Conditional expectation in an operator algebra. Tohoku Math. J. **6**, 177–181 (1954)
35. L. Accardi, A. Frigerio, J.T. Lewis, Quantum stochastic processes. Publ. Res. Inst. Math. Sci. **18**, 97–133 (1982)
36. V.P. Belavkin, Reconstruction theorem for quantum stochastic processes. Theor. Math. Phys. **3**, 409–431 (1985). arXiv:math/0512410
37. R. Haag, D. Kastler, An algebraic approach to quantum field theory. J. Math. Phys. **5**, 848–861 (1964)
38. K. Kraus, General state changes in quantum theory. Ann. Phys. **64**, 311–335 (1971)
39. G. Ludwig, Attempt of an axiomatic foundation of quantum mechanics and more general theories, II, Commun. Math. Phys. **4**, 331–348 (1967); Attempt of an axiomatic foundation of quantum mechanics and more general theories, III, ibid **9**, 1–12 (1968)
40. J. Schwinger, The algebra of microscopic measurement, Proc. Nat. Acad. Sci. U.S. **45**, 1542–1554 (1959); The geometry of quantum states ibid. **46**, 257–265 (1960)
41. K. Kraus, *States, Effects, and Operations: Fundamental Notions of Quantum Theory*. Lecture Notes in Physics, vol. 190 (Springer, Berlin, 1983)
42. C. Anantharaman-Delaroche, J.F. Havet, On approximate factorizations of completely positive maps. J. Func. Anal. **90**, 411–428 (1990)
43. J.A. Mingo, The correspondence associated to an inner completely positive map. Math. Ann. **284**, 121–135 (1989)
44. C. Anantharaman-Delaroche, Amenable correspondences and approximation properties for von Neumann algebras. Pac. J. Math. **171**, 309–341 (1995)

45. J.A. Mingo, Weak containment of correspondences and approximate factorization of completely positive maps. J. Func. Anal. **89**, 90–105 (1990)
46. D. Buchholz, C. D'Antoni, K. Fredenhagen, The universal structure of local algebras. Commun. Math. Phys. **111**, 123–135 (1987)
47. W. Arveson, Subalgebras of C*-algebras. Acta Math. **123**, 141–224 (1969)
48. A. Barchielli, V.P. Belavkin, Measurements continuous in time and a posteriori states in quantum mechanics. J. Phys. A: Math. Gen. **24**, 1495–1514 (1991)

# Chapter 6
# Response Theory Supporting Dressed Photons

**Itsuki Banno**

**Abstract** For a theoretical description of the dressed photon (DP) introduced in Chap. 1, it is essential to develop a novel method different from those applied to conventional optics. The difficulties lie in the facts that the familiar concepts such as the electric field, electric permittivity, and polarization become irrational in near-field optics (NFO), and that NFO is inevitably related to the many-electron problem. Starting from the redefinition the DP in a general manner, this chapter is devoted to explain above mentioned difficulties and develops the linear response theory applicable to the optical near field and the DP. The derived linear single susceptibility relates the scalar and vector potentials (as the cause) to the induced charge and current densities (as the result), guarantees the gauge invariance and the charge conservation law, and is soundly applicable to the many-electron system in NFO as the replacement of ordinary two susceptibilities, namely, the electric permittivity and magnetic permeability that relate the electric and magnetic fields (as the cause) to the polarization and magnetization (as the result), respectively. The present formulation will be extensible to the nonlinear single susceptibility, which is essential for a future challenge to describe the DP.

## 6.1 Introduction

A series of experiments in near-field optics (NFO) utilizing non-metallic materials was performed under non-resonant conditions and/or forbidden-transition conditions, and showed various phenomena which have not been observed in conventional optics. Such the NF optical phenomena are:

---

I. Banno (✉)
Graduate Faculty of Interdisciplinary Research Faculty of Engineering,
University of Yamanashi, 4-3-11 Takeda, Kofu, Yamanashi 400-8511, Japan
e-mail: banno@yamanashi.ac.jp



T. Yatsui (ed.), *Progress in Nanophotonics 5*, Nano-Optics and Nanophotonics,
https://doi.org/10.1007/978-3-319-98267-0_6

1. photo-chemical reaction with insufficient photon energy (photo-chemical vapor deposition [1], lithography [2], and etching [3]),
2. frequency up-conversion using nano-scaled dye grain [4, 5],
3. non-adiabatic transition leading to energy concentration among semiconductor nanodots (nanofountain) [6] and nanophotonic device [7],
4. high-efficient light-emitting diode and laser using indirect-transition type semiconductor,e.g., silicon [8–10], and
5. giant magneto-optical effect that brings about in the device fabricated in the same manner as the above light-emitting diodes [10–12].

For comprehensive review of the experiments, see the previous chapter and [10, 13]. These NF specific transitions may be achieved owing to adequate compensation of energy and/or other quantity supplied from the responding materials. To allow such transitions theoretically, the concept Dressed Photon and Dressed-Photon–Phonon are suggested; those are the EM field localized around electron excitation on a nanometric scale and easily interacting with phonons in the responding material [13–15].

This chapter redefines the *Dressed Photon (DP)* in a general manner, and develops the response theory, which is adequate for NFO and applicable to the DP. Reconsideration on some fundamental issues are of necessary:

- The concept of electric field is not available to describe the NF optical system under non-resonant condition. Instead, the scalar and vector potentials are essential.
- In such the NF optical system, therefore, it is a logical fallacy to apply the macroscopic constitutive equations (MCEs) in terms of electric field and permittivity.
- The NFO is inevitably related to many-electron problem via the longitudinal electric field, which characterize NF optical system. Actually, in ordinary Hamiltonian, the longitudinal electric field is converted to the two-body Coulomb interaction, so that the analysis of the NF optical effect is difficult. A novel theory from the scratch is necessary.
- Considering that the Coulomb interaction is the origin of diversity of materials the engineering of the longitudinal electric field in NFO possibly leads to new paradigm; this is a future challenge.

The contents in this chapter are as follows: Section 6.2 redefines the DP in a general viewpoint based on the Heisenberg's uncertainty principle. Section 6.3 characterizes and classifies NF optical system, and the DP system. Furthermore, the irrationality of the concept "electric field" in NFO is pointed out based on *the non-relativistic nature* of the system. Section 6.4 presents the strategy and requisites for the response theory applicable to the optical near field and the DP. In Sect. 6.5, the linear response theory is applied to an one-electron model in NFO to highlight the non-resonant condition imposed on non-metallic material; also to examine the effective condition for the MCEs familiar in conventional optics. Section 6.6 develops the response theory equally associated with the scalar and vector potentials in the NF optical many-electron system. The functional derivatives of the action integral by means of the electromagnetic (EM) potential derives the linear and nonlinear single susceptibilities, that guarantee the charge conservation law and the gauge invariance. In particular, the nonlinear susceptibility is essential for a future challenge to describe the DP. Section 6.7 gives a summary of this chapter. The present approach is on the halfway to the new paradigm, where the DP plays an important role, and a further conceptual development is of necessary.

## 6.2 Tentative Definition of Dressed Photon

The experiments presented in the beginning of this chapter were performed under certain inhibitory conditions, namely, the non-resonant condition, indirect-transition condition, dipole-forbidden-transition condition, and spin-forbidden-transition condition. Under these conditions, the incident EM field carries physical quantities mismatched for the electronic excitation, namely, the mismatched energy, momentum, orbital angular momentum or spatial parity, spin angular momentum, respectively. The NF specific transitions seem to be achieved owing to adequate compensation associated with the responding materials.

Actually, if a localized EM field exists, it essentially interacts with the responding material, which possibly provides the compensation energy and/or other quantity needed for the electronic transition. This fact is explained by the Heisenberg's uncertainty principle without knowing the details of the interaction.

Firstly, let us consider the EM field in a macroscopic-scale material. the EM field satisfies the dispersion relationship between the energy, $E$, (or angular frequency) and the momentum, $p$ (or wavenumber); this relationship is derived from the condition that the field can propagate for a long range. Such a field is reference field, so-called on-shell field.

If one notice an EM field localized in a small-scale material, for examples, a nanostructure in the semiconductor or the probe-tip of NF optical microscope, such the EM field possesses the uncertainty $\Delta p \sim \hbar/\Delta x$, where $\Delta x$ is the representative size of the small-scale material. Due to the uncertainty, $\Delta p$, the dispersion relationship becomes broad one, that means the localized field possesses off-shell component. The discrepancy of the energy and/or momentum between off-shell and on-shell component should be compensated via the responding material.

In short, if one notice a localized EM field, it essentially interacts with the responding material. Conversely, the interaction with the responding material supports the localization of the EM field. Therefore, the EM field localized on a small scale is qualitatively different from the reference EM field in a large-scale material, and is suitable to be named as *dressed photon* (DP). Therefore, a tentative definition of the DP is given as the EM field localized on a small scale and inevitably interacts with the responding material.

The present redefinition corresponds to *the existence theorem of the DP* without mentioning about the interaction with the responding material, and is regarded as a generalization of the original concepts, Dressed Photon and/or Dressed Photon–Phonon, suggested in [13–15], that is, the localized EM field has a wide range in wavenumber, so that interacts with the responding material.

The above discussion is about the exchange of momentum and associated energy with the responding material based on the Heisenberg uncertainty principle, therefore, such the exchange is caused by the non-commutativity of the canonial coordinate and momentum. From a general viewpoint, the non-commutative algebra associated with the NF-optical system enables the exchange of the corresponding physical quantity between the EM field of interest and the responding material. From another viewpoint, the interaction with the responding material seems to originate in the translational symmetry breakdown. If so, the rotational symmetry breakdown possibly brings about the exchange of the orbital and/or spin angular momentum with the responding material. In general, the symmetry breakdown enables the more number of states to commit with the interaction. It is a future consideration that the non-commutativity of algebra and the activation of states due to the symmetry breakdown bring about what quantity to be exchanged, and how much possible to exchange.

## 6.3 Characterization of NF Optical Systems

The DP appears in a NF optical system, i.e., in the vicinity of nanostructure. In this section, the NF optical system is characterize and classified, comparing with conventional optical system. An irrationality of the concept of "electric field" in a NF optical system is pointed out based on *the non-relativistic nature* of the system.

### *6.3.1 Longitudinal and Transverse Electric Fields in NFO*

To make clear a difference between NF optical and conventional optical systems, suppose a material is placed in the vicinity of the nanostructure, which works as a light source (Fig. 6.1). For a while, the material is assumed to be on a small scale for simplicity, although the size of the responding electron system is not restricted to be small.

In such a system under the NF incidence condition, the responding material is exposed to the longitudinal and transverse electric fields simultaneously, while in a system under the far-field incidence condition, the responding material is exposed only to the transverse one, which survives far from the light source. Therefore, the coexistence of longitudinal and transverse electric fields distinguishes such the system under NF incidence condition from that under the far-field incidence condition.

Here, the longitudinal electric field is originates from the charge density on the nanostructure, obeys Coulomb's law, and has non-radiative nature to localize around the nanostructure. On the other hand, the transverse electric field originates from the transverse current density on the nanostructure, obeys Ampere-Maxwell's law and Faraday's law, and has radiative nature to propagate far from the light source, accompanied by the magnetic field; see Table 6.1 for the summary. (The longitudinal

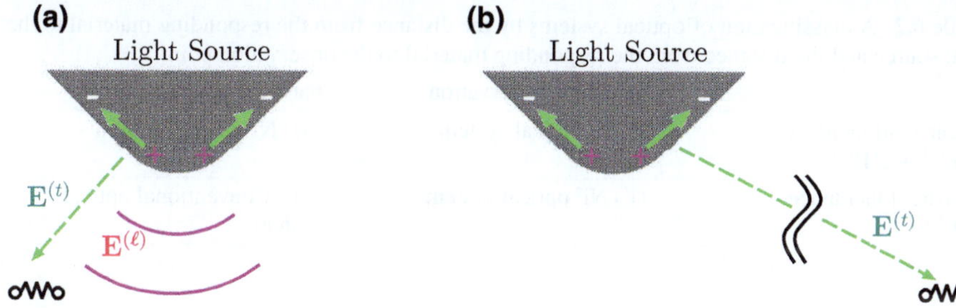


**Fig. 6.1** The responding materials under near- and far-field incidence conditions: **a** The former is exposed to the incident longitudinal and transverse electric fields simultaneously, **b** while the latter is exposed to only the transverse one

**Table 6.1** Comparison between the longitudinal and transverse electric fields, $\mathbf{E}^{(\ell)}$ and $\mathbf{E}^{(t)}$. The longitudinal current density, $\mathbf{j}^{(\ell)}$ is not an independent source because it is related to the charge density, $\rho$ via the charge conservation law, $\partial_t \rho + \nabla \cdot \mathbf{j}^{(\ell)} = 0$

| | $\mathbf{E}^{(\ell)}$ | $\mathbf{E}^{(t)}$ |
|---|---|---|
| Source | $\rho$ | $\mathbf{j}^{(t)}$ |
| Governing law | $\nabla \cdot \mathbf{E}^{(\ell)} = \dfrac{\rho}{\epsilon_0}$ | $\begin{cases} \nabla \times \mathbf{E}^{(t)} + \partial_t \mathbf{B} = \mathbf{0} \\ \nabla \times \mathbf{B} - \frac{1}{c^2}\partial_t \mathbf{E}^{(t)} = \frac{\mathbf{j}^{(t)}}{\epsilon_0 c^2} \end{cases}$ |
| Propagation nature | Non-radiative | Radiative |

current density is determined via the charge conservation law, once the charge density is known, and that is not an independent source.) Therefore, under the NF incidence condition, the coexisting two types of incidences in the NF optical system have distinct properties, and should be treated separately as the the cause of response.

### 6.3.2 *Irrationality of Electric Field in NFO*

Furthermore, owing to *the non-relativistic nature* of the system, the scalar and vector potentials appear in a different manner in the Hamiltonian, which governs the electron response, e.g., (6.13) in Sect. 6.5.1 for the spinless one-electron system. Considering that the scalar and vector potentials under the Coulomb gauge represent the longitudinal and transverse electric fields, respectively, one may confirm that the two types of incidences in NFO possibly cause different responses. This fact represents for the irrationality of the concept *electric field* under the NF incidence condition. In Sect. 6.5, it will be shown that this irrationality comes into observation only in the NF system (I) in Table 6.2 under a non-resonant condition; such optical system is complement to the familiar systems in conventional optics.

**Table 6.2** A classification of optical systems by the distance from the responding material to the light source and the distance from the responding material to the observation point

| | Near-field observation | Far-field observation |
|---|---|---|
| Near-field incidence: $\Delta\mathbf{E}^{(\ell)} + \Delta\mathbf{E}^{(t)}$ | (I) NF optical system | (II) NF optical system |
| Far-field incidence: $\Delta\mathbf{E}^{(t)}$ | (I′) NF optical system | (II′) Conventional optical system |

### 6.3.3 *Classification of Optical Systems*

Let us classify optical systems. The two systems under the near- and far-field incidence conditions in Fig. 6.1 are subdivided into two classes depending on the near- or far-field observation condition. These four classes are listed in Table 6.2. In particular, the systems of (I′) and (II′) are the limiting cases of null longitudinal incidence of the systems (I) and (II), respectively. So, in the systems of (I′) and (II′), the longitudinal response vanishes and the difference of responses may not observed. In the following, therefore, we concentrate our main interest in the systems (I) and (II), in which the longitudinal incidence exists.

In practice, the light source is something like a nanostructure irradiated by the external light, and the light sink (observation detector) is like a probe-tip of the NF optical microscope or a molecule to be dissociated, and those are idealized as follows: The light source is robust and not affected by the scattered EM field from the responding material, and the light sink is complete absorber and not affects to the responding material. The idealized description by a response theory is available under such a situation that the multiple-scatterings between the responding material and both the light source and sink are negligible.

### 6.3.4 *DP and Optical Near Field*

So far as mentioned above, the responding electron system interacting with the localized EM field is assumed to be on a small scale. However, this situation is not always true. For example, in the systems of the cases "4." and "5." in the experiments introduced in Sect. 6.1, the carrier electron and hole possibly distribute over a wide range in the semiconductor. The light source is considered as the event of electron-hole recombination to produce the incident EM field around a specific place (considered as a phonon cavity at the present moment [16–18]), and the light sink is the event to radiate to the vacuum at the same place after the incident EM field interacts with the responding material to form the DP. If this picture is true, such the system associated with the DP is regarded as a limit of the NF optical system (I), and approach employing the response theory is available. Therefore, the size of the responding material, is not restricted to be small in the followings.

## 6.4 Strategy of Approach to Dressed Photon

In Sect. 6.2, one may confirm the existence of the DP without knowing the interaction with the responding material, and, in Sect. 6.3, one may comprehend that the NF optical system, in which the DP appears, is qualitatively different from conventional optical system. The next question and our main interest is: "What interaction support the existence of the DP ?" or "How to describe the physics of the DP ?" In this section, the strategy of approach to the DP employing the response theory is explained.

### *6.4.1 Thomas–Fermi Theory: A Pedagogical Guide*

The Thomas–Fermi theory is a pedagogical guide to a comprehensive understanding of the present approach developed in the following sections, although here treats a metallic systems in stead of the non-metallic system of our interest.

Suppose a local charged impurity, $\Delta\rho_{\rm ext}(\mathbf{r})$, is introduced in the metallic material on a large scale; this impurity provides scalar potential incidence. We like to know the scalar potential change, $\Delta\phi(\mathbf{r})$ considering the effect of the induced electron charge density, $\Delta\rho(\mathbf{r})$ in the responding metallic material, of which Fh size is assumed to be rather large scale. The light sink is unphysically not set but it is implicitly assumed that the scalar potential is measurable as it is.

The Thomas–Fermi approach to this electrostatic problem is attributed to solve the coupling equations of the Poisson equation and the constitutive equation. The Poisson equation relates the scalar potential under the Coulomb gauge (*as the result*) and the charge density (*as the cause*):

$$-\triangle(\Delta\phi(\mathbf{r})) = \frac{1}{\epsilon_0}\left(\Delta\rho_{\rm ext}(\mathbf{r}) + \Delta\rho(\mathbf{r})\right) . \tag{6.1}$$

The constitutive equation relates the charge density (*as the result*) and the scalar potential (*as the cause*) in accordance with the classical or quantum mechanics. A simple derivation of the constitutive equation in Thomas–Fermi theory is as follows [19]. The change of the scalar potential, $\Delta\phi(\mathbf{r})$, near the impurity motivate to change the local Fermi energy, $\varepsilon_{\rm f}(\mathbf{r})$, by the amount, $q\Delta\phi(\mathbf{r})$, where $q$ is the electron charge. But $\varepsilon_{\rm f}$ is kept constant owing to the large scale of the responding metallic material. This situation is achieved by the change of electron density, $\Delta\rho(\mathbf{r})$, contributed from the states near the Fermi energy with the energy width, $q\Delta\phi(\mathbf{r})$. Such the the change of electron density is as follows:

$$\Delta\rho(\mathbf{r}) = -q^2\mathcal{D}(\varepsilon_{\rm f})\Delta\phi(\mathbf{r}) , \tag{6.2}$$

where $\mathcal{D}(\varepsilon_f)$ is density of states in the vicinity of the Fermi energy, i.e., the states' number per unit energy and unit volume. This constitutive equation, (6.2), is the linearized version of that in the original Thomas–Fermi theory.

The elimination of the charge density from these two equations (6.1) and (6.2) leads to the equation for the scalar potential of our interest:

$$-\Delta(\Delta\phi(\mathbf{r})) = \frac{1}{\epsilon_0}\left(\Delta\rho_{\text{ext}}(\mathbf{r}) + \Delta\rho[\Delta\phi](\mathbf{r})\right), \tag{6.3}$$

$$\therefore (-\Delta + \lambda_{\text{TF}}^2)\Delta\phi(\mathbf{r}) = \frac{1}{\epsilon_0}\Delta\rho_{\text{ext}}(\mathbf{r}), \tag{6.4}$$

$$\text{where } \lambda_{\text{TF}} = |q|\sqrt{\frac{\mathcal{D}(\varepsilon_f)}{\epsilon_0}}.$$

The parameter, $\lambda_{\text{TF}}$ in (6.5) describes the localization or the screening of the scalar potential change caused by the impurity charge density and the induced charge density in the responding metallic material. Actually, in the case of $\Delta\rho_{\text{ext}} = \Delta q_{\text{ext}}\delta^3(\mathbf{0})$, the solution of (6.5) is the Yukawa-type function,

$$\Delta\phi(\mathbf{r}) = \frac{\Delta q_{\text{ext}}}{4\pi\epsilon_0}\frac{e^{-\lambda_{\text{TF}}r}}{r} \tag{6.5}$$

As described above, the constitutive equation is responsible for the induced electron charge density caused in the material, which modifies the scalar potential yielded by the bare charged impurity. The present description seems to fit well with the DP, so that, the fundamental idea is generalized in the next subsection.

### 6.4.2 *Single Susceptibility*

The response function theory determines the constitutive equation in terms of susceptibility that are derived from the elementary law in classical or quantum mechanics.

Optical system is governed by the coupled equations of the EM field (the scalar and vector potentials; $A^\nu(x)$) and the source field (charge and current field densities; $j^\mu(x)$), that is, Maxwell's wave equation and Heisenberg's equation[1];

$$\left(\delta^\mu_\nu \Box - \partial^\mu\partial_\nu\right) A^\nu(x) = \frac{1}{\epsilon_0 c} j^\mu(x), \tag{6.6}$$

$$\frac{\partial}{\partial t} j^\mu(x) = \left[j^\mu(x), H\right], \tag{6.7}$$

[1]We use four-element representation with Einstein's rule; see Sect. 6.6 for detials.

where $H$ is the Hamiltonian and $[\,\cdots\,]$ is the Poisson bracket in classical treatment and is replaced by $\frac{1}{i\hbar}[\,\cdots\,]$ in quantum treatment. The solution of the Maxwell's wave equation (6.6) is given as the scalar and vector potentials as the functional of the charge and current density, $A^{\mu}[j^{\nu}]$ and that of the Heisenberg equation (6.7) is vise versa, $j^{\nu}[A^{\mu}]$, which is so-called constitutive equation. Once one obtains the constitutive equation, eliminating $j^{\nu}(x)$ from (6.6) and (6.7) results in the equation for $A^{\mu}(x)$ modified by the interaction with the responding material:

$$\left(\delta^{\mu}{}_{\nu}\Box - \partial^{\mu}\partial_{\nu}\right) A^{\nu}(x) = \frac{1}{\epsilon_0 c} j^{\mu}[A^{\varsigma}](x)\,, \tag{6.8}$$

or for the changed portion,

$$\left(\delta^{\mu}{}_{\nu}\Box - \partial^{\mu}\partial_{\nu}\right) \Delta A^{\nu}(x) = \frac{1}{\epsilon_0 c} \Delta j^{\mu}[\Delta A^{\varsigma}](x)\,. \tag{6.9}$$

This equation describes the EM field modified by the interaction with the material, corresponding to (6.3) in the Thomas–Fermi theory, and possibly describe the DP.

In practice, (6.6) is linear differential equation with inhomogeneity and can be solved rigorously, while (6.7) is difficult to solve, in general, so that one often employs a perturbation method, i.e., the linear and nonlinear response theory and determine the induced current density in the righthand side of (6.8). In the approach employing the response theory, the induced source field (the difference from the reference source field in the non-perturbed system), $\Delta j^{\nu}(x)$ is described by the total derivative with respect to the perturbed EM field $\Delta A^{\mu}(x_1)$ and the partial derivative coefficient, $\chi^{\nu}_{\mu}(x, x_1)$ is the susceptibility (response function), e.g., in the linear response theory,

$$\Delta j^{\nu}(x) = \int \mathrm{d}^4 x_1\, \chi^{\nu}{}_{\mu}(x, x_1) \Delta A^{\mu}(x_1). \tag{6.10}$$

Remark that the total derivative means the collection of all the contribution from the cause, so that is described by means of the convolution of the vector index and the coordinate of the perturbation EM field. Therefore, the susceptibility is non-local and non-diagonal, in general.

The susceptibility relates the induced current density (*as the result of response*) and the EM potential (*as the cause of response*). This type of susceptibility is named as *single susceptibility*, comparing with the two susceptibilities, namely, the electric permittivity and magnetic permeability in the MCEs.

The single susceptibility by itself should be responsible for the charge conservation law and gauge invariance. Because such the susceptibility may be determined without any subsidiary condition starting from the Heisenberg's equation, which maintains the charge conservation law and the gauge invariance.

### 6.4.3 Irrationality of Polarization and MCEs

As explained in the previous subsection, the fundamental constitutive equation relates between the EM potential (as the cause of response) and the induced charge and current densities (as the result of response). This manner is contrasting with the MCEs, which is the relationship between the electric and magnetic fields (*as the case of response*) and the polarization and magnetization (*as the result of response*).

The concepts of polarization and magnetization were introduced before the invention of the Dirac's delta function, and were utilized to avoid the delta-function like singularity in the charge and current densities induced on an interface. Actually, replacing the effect of such singular source with the Maxwell's boundary conditions, the singularity is settled within that of the Heaviside's step function. However, as one may notice after careful derivation of the Maxwell's boundary conditions, there exists a prerequisite that the interface is flat over the range of a light wavelength. Therefore, Maxwell's boundary condition is not adequate to treat the optical near field of a nanostructure. Under the far-field observation, the structure on a nano-metric scale is averaged out and the treatment based on MCEs are justified to some degrees as described in Sect. 6.5.2. However, in the context of NFO and the DP, there are some logical fallacies of the MCEs in terms of the polarization and magnetization:

- As pointed out in Sect. 6.3.2, and will be shown details in Sect. 6.5.4, the total electric field may not be regarded as the cause of response in a certain NF optical system. Therefore, the MCEs may not describe such a NF optical system, and the single susceptibility is essential. At least, the longitudinal and transverse electric field should be treated separately as the cause of response under the long wavelength approximation (LWA); see Sect. 6.5.
- The fundamental degrees of freedom of the source is the charge density and the transverse current density, which carry totally three components. (The longitudinal current density is excluded because it is determined via the charge conservation law, once the charge density is known.) On the other hand, the source in the MCEs are the polarization and magnetization, which carry totally six components and should include redundant degrees of freedom. Therefore, the MCEs implicitly includes constraint for the unphysical redundancy.
- Another irrationality of the MCEs in a chiral optical systems was pointed out by K. Cho [20, 21]. He derived a Taylor expansion for his single susceptibility [22], using the LWA ($ka \ll 1$, where $2\pi/k$ is the light wavelength and $a$ is the representative size of the material) and has shown the term of leading order, $\mathcal{O}(ka)^0$, gives the electric permittivity, and the term of order $\mathcal{O}(ka)^2$ gives the magnetic permeability. This separability of the electric and magnetic responses, however, break down in a chiral system due to the mixing of the electric and magnetic response in the order $\mathcal{O}(ka)^1$.

As a result, in NFO, the single susceptibility is essential, instead of the electric permittivity and magnetic permeability in the MCEs.

### 6.4.4 NFO and Many-electron Problem

Why the longitudinal electric field and a comparison between the responses to the two types of electric fields have not been attracted in NF optical theory? Perhaps, the reason is the procedure to construct the ordinary Hamiltonian for a many-electron system, in which the longitudinal electric field is rewritten to the two-body Coulomb interaction and eliminated. To do so, the quantum many-electron effect (so-called exchange-correlation effect) is considered to make the ground and excited states bounded. The transverse electric field is remained as the cause of response, so that this Hamiltonian is compatible with conventional optics, however, inadequate for NFO because this Hamiltonian unequally treats the incident longitudinal and transverse electric fields. Furthermore, with this Hamiltonian, the response to the longitudinal electric field incidence gets into the Coulomb interaction and is difficult to analyze; the Coulomb interaction itself is difficult to be treated, and often ignored innocently. In short, the NFO is inevitably related to the many-electron problem via the longitudinal electric field, although this fact has not been well recognized for a long time.

An equal treatment of the scalar and vector potentials as the coexisting cause of response in the NF optical system should be achieved, overcoming the many-electron problem. That is, one should separate the effect of the external scalar potential incidence from the the effect of the internally existing scalar potential; the latter is the quantum many electron effect (exchange-correlation effect) which make the ground state bounded.

### 6.4.5 Requisites for the Response Theory in NFO

Here is the summary of the requisites for the response theory to describe physics in NFO including the DP:

1. A single susceptibility is essential, instead of the electric permittivity and the magnetic permeability used in the MCEs; see Sects. 6.3.2 and 6.4.3. The single susceptibility relates the scalar and vector potentials (as the cause) to the charge and current densities (as the result).
2. The scalar potential as the cause of response should be separated from the inherently existing scalar potential yielded by the charge of electron in many-electron system. The ordinary Hamiltonian is inadequate for the NF optical system because the two portion is converted together to the Coulomb interaction. Therefore, a novel formulation from the scratch is of necessary.
3. The susceptibility by itself should be responsible for the charge conservation law and the gauge invariance; see Sect. 6.4.2.
4. The nonlinear response is essential to describe the DP, because it is located extremely close to or embedded in the electron system. Systematic treatment of nonlinear susceptibility is desirable.

5. A formulation in the Heisenberg picture is preferable; it enable the classical-quantum correspondence easily and assign what is the quantum effect. Furthermore, it is essential to separate the operators and states to recover the microscopic quantities from macroscopically observed quantities, with the help of the micro-macro duality developed by I. Ojima [23]; this is a future challenge.
6. The light source and sink is assumed to bring about no multiple scattering with the responding material. This is a prerequisite for the description using a response theory; see Sect. 6.3.3.

## 6.5 Linear Response Theory of a Simplified NF Optical System

Before developing a general formulation of the response theory to describe the optical near field and the DP, this section reveals the irrationality of the electric field in a NF optical system, and emphasizes that the NF optical system is qualitatively different from conventional optics.

### *6.5.1 Linear Response of One-electron System*

Suppose a small-scale material is exposed to the incident scalar and vector potentials (under the Coulomb gage), $\Delta\phi(\mathbf{r}, t)$ and $\Delta\mathbf{A}(\mathbf{r}, t)$, which represent for the longitudinal and transverse electric fields, $\Delta\mathbf{E}^{(\ell)}(\mathbf{r}, t) = -\nabla(\Delta\phi(\mathbf{r}, t))$ and $\Delta\mathbf{E}^{(t)}(\mathbf{r}, t) = -\frac{\partial}{\partial t}\Delta\mathbf{A}(\mathbf{r}, t)$, respectively. The incident scalar and vector potentials, $\Delta\phi(\mathbf{r}, t)$ and $\Delta A_i(\mathbf{r}, t)$, are assumed to be monochromatic with the angular momentum $\omega$, and possess the wavelength or the spatial variation on a large scale relative to the size of the responding material, so that, the LWA is applied under the Coulomb gauge.

$$\Delta\phi(\mathbf{r}, t) = \Delta\phi(\mathbf{r}) \cos\omega t = \big(\Delta\phi(\mathbf{0}) - \Delta\mathbf{E}^{(\ell)}(\mathbf{0}) \cdot \mathbf{r}\big) \cos\omega t \,, \tag{6.11}$$

$$\Delta\mathbf{A}(\mathbf{r}, t) = \Delta\mathbf{A}(\mathbf{r}) \sin(\omega t + \xi) = -\frac{1}{\omega}\Delta\mathbf{E}^{(t)}(\mathbf{0}) \sin(\omega t + \xi) \,, \tag{6.12}$$

where $\mathbf{0}$ is the representative point of the responding material and $\xi$ is the phase difference between the two incident potentials.

Instead of the polarization and magnetization, we employ the induced charge and current densities, $\Delta\rho(\mathbf{r}, t)$ and $\Delta\mathbf{j}(\mathbf{r}, t)$, as the fundamental source (*the results of response*) to examine the distinct responses to the longitudinal and transverse electric fields (*the causes of response*), $\Delta\mathbf{E}^{(\ell)}(\mathbf{0}, t)$ and $\Delta\mathbf{E}^{(t)}(\mathbf{0}, t)$, where $\mathbf{0}$ is the representative position in the electron system.

Using spinless one-electron system, let us evaluate the induced charge and current densities caused by the coexisting incidences of the scalar and vector potentials. The total Hamiltonian is:

$$\hat{H} = \frac{1}{2m}\left(\frac{\hbar}{i}\frac{\partial}{\partial x_i(t)} - qA_i(\mathbf{x}(t),t)\right)\left(\frac{\hbar}{i}\frac{\partial}{\partial x_i(t)} - qA_i(\mathbf{x}(t),t)\right) + q\phi(\mathbf{x}(t),t)\,, \tag{6.13}$$

where $t$ is time, $\mathbf{x}(t)$ is the position of the electron, $q(=-e), m$ are the electron charge and mass, respectively.[2] The perturbation Hamiltonian is:

$$\int \mathrm{d}^3r\,\left(\hat{\rho}(\mathbf{r},t)\Delta\phi(\mathbf{r},t) - \hat{j}_i(\mathbf{r},t)\Delta A_i(\mathbf{r},t)\right), \tag{6.14}$$

where $\hat{\rho}(\mathbf{r},t)$, $\hat{j}_i(\mathbf{r},t)$ are the Heisenberg operators of the charge and current densities defined as:

$$\hat{\rho}(\mathbf{r},t) = q\delta^3(\mathbf{r}-\mathbf{x}(t))\,,$$

$$\hat{j}_i(\mathbf{r},t) = \frac{q}{2m}\left\{\left(\frac{\hbar}{i}\frac{\partial}{\partial x_i(t)} - qA_i(\mathbf{x}(t),t)\right)\delta^3(\mathbf{r}-\mathbf{x}(t))\right. \tag{6.15}$$

$$\left. +\delta^3(\mathbf{r}-\mathbf{x}(t))\left(\frac{\hbar}{i}\frac{\partial}{\partial x_i(t)} - qA_i(\mathbf{x}(t),t)\right)\right\}. \tag{6.16}$$

The linear response theory leads to the operators of the induced charge and current densities as follows:

$$\Delta\hat{\rho}(\mathbf{r},t) = \int_{-\infty}^{t}\mathrm{d}t_1\int\mathrm{d}^3r_1\left\{\frac{1}{i\hbar}\left[\hat{\rho}^{(0)}(\mathbf{r},t)\,,\,\hat{\rho}^{(0)}(\mathbf{r}_1,t_1)\right]\Delta\phi(\mathbf{r}_1,t_1)\right.$$
$$\left.-\frac{1}{i\hbar}\left[\hat{\rho}^{(0)}(\mathbf{r},t)\,,\,\hat{j}^{(0)}_{i_1}(\mathbf{r}_1,t_1)\right]\Delta A_{i_1}(\mathbf{r}_1,t_1)\right\}, \tag{6.17}$$

$$\Delta\hat{j}_i(\mathbf{r},t) = \int_{-\infty}^{t}\mathrm{d}t_1\int\mathrm{d}^3r_1\left\{\frac{1}{i\hbar}\left[\hat{j}^{(0)}_i(\mathbf{r},t)\,,\,\hat{\rho}^{(0)}(\mathbf{r}_1,t_1)\right]\Delta\phi(\mathbf{r}_1,t_1)\right.$$
$$\left.-\frac{1}{i\hbar}\left[\hat{j}^{(0)}_i(\mathbf{r},t)\,,\,\hat{j}^{(0)}_{i_1}(\mathbf{r}_1,t_1)\right]\Delta A_{i_1}(\mathbf{r}_1,t_1)\right\}$$
$$-\frac{q}{m}\hat{\rho}^{(0)}(\mathbf{r},t)\Delta A_i(\mathbf{r},t)\,, \tag{6.18}$$

where $\hat{\rho}^{(0)}$ and $\hat{\mathbf{j}}^{(0)}$ are the charge and current density operators, respectively, in the non-perturbed system. The last term in (6.18) is originated from *the non-relativistic nature* of the system and needed to maintain the charge conservation law.

We take the expectation values of (6.17) and (6.18) using the ground state, $\varphi_0(\mathbf{r})$, and inserts the projection operator [the lefthand side of the second equation of (6.19)],

---

[2]The Einstein's rule is used for the summation over the vector indices, e.g., $A_iA_i = \sum_{i=1}^{3} A_iA_i$ .

assuming that the two orbitals are real function and form the normalized orthogonal complete set:

$$\int d^3r\, \varphi_m(\mathbf{r})\varphi_n(\mathbf{r}) = \delta_{m\,n}\,, \qquad \sum_m \varphi_m(\mathbf{r})\varphi_m(\mathbf{r}') = \delta^3(\mathbf{r}-\mathbf{r}')\,, \tag{6.19}$$

where $\varphi_m(\mathbf{r})$ satisfies,

$$\hat{H}^{(0)}\varphi_m(\mathbf{r}) = \hbar\omega_m\,\varphi_m(\mathbf{r})\,, \quad (m=0,1)\,. \tag{6.20}$$

The excitation energy is $\hbar\Delta\omega_1 \equiv \hbar\omega_1 - \hbar\omega_0 > 0$; this finite excitation energy represents that the responding material is non-metallic, such as a molecule, a nano-structured semiconductor and insulator. The real orbitals carry even temporal parity so that there is null magnetic field or null vector potential in the non-perturbed Hamiltonian.

Then, substituting (6.11) and (6.12) lead to the expression for the induced charge and current densities (the results) as the total derivative with respect to the longitudinal and transverse electric fields (the causes); the partial derivative coefficients, $\chi_{\cdots}^{\cdots}(\mathbf{r},\omega)$'s are susceptibilities (response functions) that relates the causes and the results:

$$\Delta\rho(\mathbf{r},t) = \chi_j^{\rho\leftarrow(\ell)}(\mathbf{r},\omega)\,\Delta E_j^{(\ell)}(\mathbf{0},t) + \chi_j^{\rho\leftarrow(t)}(\mathbf{r},\omega)\,\Delta E_j^{(t)}(\mathbf{0},t)\,, \tag{6.21}$$

$$\Delta j_i(\mathbf{r},t) = \chi_{ij}^{\mathbf{j}\leftarrow(\ell)}(\mathbf{r},\omega)\,\Delta\dot{E}_j^{(\ell)}(\mathbf{0},t) + \chi_{ij}^{\mathbf{j}\leftarrow(t)}(\mathbf{r},\omega)\,\Delta\dot{E}_j^{(t)}(\mathbf{0},t)\,, \tag{6.22}$$

where

$$\Delta E_j^{(\ell)}(\mathbf{0},t) \equiv \Delta E_j^{(\ell)}(\mathbf{0})\cos\omega t\,, \quad \Delta E_j^{(t)}(\mathbf{0},t) \equiv \Delta E_j^{(t)}(\mathbf{0})\cos(\omega t+\xi)\,, \tag{6.23}$$

$$\Delta\dot{E}_j^{(\ell)}(\mathbf{0},t) \equiv \frac{\partial}{\partial t}\Delta E_j^{(\ell)}(\mathbf{0},t)\,, \qquad \Delta\dot{E}_j^{(t)}(\mathbf{0},t) \equiv \frac{\partial}{\partial t}\Delta E_j^{(t)}(\mathbf{0},t)\,. \tag{6.24}$$

In (6.22), the time derivatives of the two types of electric fields, i.e., $\Delta\dot{E}_j^{(\ell)}(\mathbf{0},t)$ and $\Delta\dot{E}_j^{(t)}(\mathbf{0},t)$ are regarded as the causes, for convenience, to mach the time-dependence between the result and causes. The magnetic response vanishes in the leading order under the LWA; see [20] and the third item in Sect. 6.4.3.

Furthermore, we use the well-known linear relationship between the two types of the dipole transition matrix elements,

$$\mathcal{C}_i \equiv \int d^3r\, \left(\partial_i\varphi_1(\mathbf{r})\varphi_0(\mathbf{r}) - \varphi_1(\mathbf{r})\partial_i\varphi_0(\mathbf{r})\right) = \frac{2m}{\hbar^2}\hbar\Delta\omega_1\,\mathcal{D}_i\,. \tag{6.25}$$

Equation (6.25) is derived from the matrix element of the Heisenberg equation for dipole charge density,

$$\frac{\partial}{\partial t} r_j \hat{\rho}^{(0)}(\mathbf{r}, t) = \frac{1}{i\hbar} \left[ r_j \hat{\rho}^{(0)}(\mathbf{r}, t) \,,\ \hat{H}^{(0)} \right] , \tag{6.26}$$

with using the relationship, $\hat{\rho}^{(0)}(\mathbf{r}, t) = e^{-\frac{\hat{H}^{(0)}t}{i\hbar}} \hat{\rho}^{(0)}(\mathbf{r}, 0) e^{+\frac{\hat{H}^{(0)}t}{i\hbar}}$, and the projection operator, (6.19) and (6.20).

Then, the susceptibilities leading to the induced charge density result in:

$$\chi_j^{\rho\leftarrow(\ell)}(\mathbf{r}, \omega) = \chi_j^{\rho\leftarrow(t)}(\mathbf{r}, \omega) = 2q^2 \frac{\eta}{\eta^2 - 1} \frac{1}{\hbar\omega} \mathcal{D}_j \, \varphi_0(\mathbf{r}) \varphi_1(\mathbf{r}) \,, \tag{6.27}$$

$$\text{where} \quad \eta \equiv \frac{\hbar \Delta\omega_1}{\hbar\omega} = \frac{\text{excitation energy}}{\text{photon energy}} \,, \ \text{and} \tag{6.28}$$

$$\mathcal{D}_i \equiv \int \mathrm{d}^3 r \ \varphi_1(\mathbf{r}) \, r_i \, \varphi_0(\mathbf{r}) \,. \tag{6.29}$$

This means that the response to the longitudinal and transverse electric fields are common, so that the induced charge density (as the result) has linear relationship with the total electric field (as the cause), i.e., $\Delta\rho(\mathbf{r}, t) = \chi_j^{\rho\leftarrow(\ell)}(\mathbf{r}, \omega)(\Delta E_j^{(\ell)}(\mathbf{0}, t) + \Delta E_j^{(t)}(\mathbf{0}, t))$.

The susceptibilities leading to the induced current density are not simple. Those are result in:

$$\chi_{ij}^{\mathbf{j}\leftarrow(\ell)}(\mathbf{r}, \omega) = \frac{q^2}{m\omega^2} \frac{1}{\eta^2 - 1} \mathcal{D}_j \ (\partial_i \varphi_1(\mathbf{r}) \varphi_0(\mathbf{r}) - \varphi_1(\mathbf{r}) \partial_i \varphi_0(\mathbf{r})) \ ,$$

$$\chi_{ij}^{\mathbf{j}\leftarrow(t)}(\mathbf{r}, \omega) = \frac{q^2}{m\omega^2} \left( \frac{\eta^2}{\eta^2 - 1} \mathcal{D}_j \ (\partial_i \varphi_1(\mathbf{r}) \varphi_0(\mathbf{r}) - \varphi_1(\mathbf{r}) \partial_i \varphi_0(\mathbf{r})) \right. \tag{6.30}$$

$$\left. - \delta_{ij} \varphi_0(\mathbf{r}) \varphi_0(\mathbf{r}) \right) \,. \tag{6.31}$$

The susceptibility to the transverse electric field (6.31), is composed of two terms. The first term, namely resonant term, includes the energy denominator enhanced under the resonant condition, $\eta \simeq 1$, as in the susceptibility to the longitudinal electric field (6.30). The second term, namely non-resonant term, does not include such the resonant factor. Let us examine three cases:

### 6.5.2 *Under the Far-Field Observation Condition*

Under the far-field observation condition, the total electric field is regarded as the cause of response, and furthermore, the MCE in terms of the total electric field and the permittivity is justified. The explanation is as follows: Suppose that we would like to observe the far field yielded by the induced charge and current densities; in this situation, the responding material works as the light source. The far field is insensitive to the details of the source but dominated by the spatial average of the source. Under the LWA, such average are achieved by the spatial average of the susceptibilities, that is, the following replacements in (6.27), (6.30) and (6.31),

$$\varphi_0(\mathbf{r})\varphi_1(\mathbf{r}) \quad \longrightarrow \quad \frac{1}{\mathcal{V}}\int \mathrm{d}^3r\, \varphi_0(\mathbf{r})\varphi_1(\mathbf{r}) = 0\,, \tag{6.32}$$

$$\partial_i\varphi_1(\mathbf{r})\varphi_0(\mathbf{r}) - \varphi_1(\mathbf{r})\partial_i\varphi_0(\mathbf{r})$$
$$\longrightarrow \quad \frac{1}{\mathcal{V}}\int \mathrm{d}^3r\, \partial_i\varphi_1(\mathbf{r})\varphi_0(\mathbf{r}) - \varphi_1(\mathbf{r})\partial_i\varphi_0(\mathbf{r}) = \frac{1}{\mathcal{V}}\mathcal{C}_i\,, \tag{6.33}$$

$$\varphi_0(\mathbf{r})\varphi_0(\mathbf{r}) \quad \longrightarrow \quad \frac{1}{\mathcal{V}}\int \mathrm{d}^3r\, \varphi_0(\mathbf{r})\varphi_0(\mathbf{r}) = \frac{1}{\mathcal{V}}\,. \tag{6.34}$$

The results are:

$$\overline{\chi_j^{\rho\leftarrow(\ell)}(\mathbf{r},\omega)} = \overline{\chi_j^{\rho\leftarrow(t)}(\mathbf{r},\omega)} = 0\,, \tag{6.35}$$

$$\overline{\chi_{ij}^{\mathbf{j}\leftarrow(\ell)}(\mathbf{r},\omega)} = \overline{\chi_{ij}^{\mathbf{j}\leftarrow(t)}(\mathbf{r},\omega)} = \delta_{i\,j}\frac{q^2\hbar^2}{m\,\mathcal{V}}\frac{1}{(\hbar\Delta\omega_1)^2-(\hbar\omega)^2}\,, \tag{6.36}$$

where the overline means the spatial average and $\mathcal{V}$ is the volume of the responding material.

To derive (6.36), we additionally use the trade-off relationship between the two types of dipole transition matrix elements,

$$\mathcal{D}_i\,\mathcal{C}_j = \delta_{i\,j}\,. \tag{6.37}$$

This is effective in the two-level system with well-defined parity and derived from the quantum-mechanical commutation relationship:

$$[r_i\,,\,\frac{\hbar}{i}\partial_j] = i\hbar\,\delta_{ij}\,, \quad \text{i.e.,} \quad r_i\left(\frac{\hbar}{i}\partial_j\cdots\right) + \frac{\hbar}{-i}\partial_j\,(r_i\cdots) = i\hbar\delta_{ij}\cdots\,. \tag{6.38}$$

Inserting the projection operator between $r_i$ and $\frac{\hbar}{i}\partial_j$, and eliminating the null integrals caused by mismatched parity result in (6.37). From (6.25) and (6.37), $\mathcal{D}_i$ and $\mathcal{C}_i$ are specified as:

$$\mathcal{D}_i = \frac{1}{\mathcal{C}_i} = \frac{\hbar}{\sqrt{2m\,\hbar\Delta\omega_1}}\,. \tag{6.39}$$

Therefore, the total electric field is regarded as the cause of response under the far-field observation condition. The null response represented in (6.35) is reasonable because the induced charge density yields the longitudinal electric field, which has non-radiative nature and vanishes in the far-field regime. The spatial average of the induced current density derived from (6.36) is only the source that contribute to the far field.

**Table 6.3** The summary of the results; the validity of total electric field as the cause of response (Sects. 6.5.2–6.5.4) and the validity of the MCE in terms of the permittivity (Sect. 6.5.5)

| | **Near-field observation**<br>the source: $\Delta\rho$ and $\Delta\mathbf{j}$ | **Far-field observation**<br>the source: $\overline{\Delta\mathbf{j}}$ |
|---|---|---|
| **Near-field incidence**<br>$\Delta\mathbf{E}^{(\ell)} + \Delta\mathbf{E}^{(t)}$ | **(I) NF optical system**<br>non-resonant / resonant | **(II) NF optical system**<br>non-resonant / resonant |
| validity of the electric field | NG / OK | OK / OK |
| validity of the MCE | NG / NG | OK / OK |
| **Far-field incidence**<br>$\Delta\mathbf{E}^{(t)}$ | **(I′) NF optical system**<br>non-resonant / resonant | **(II′) Conventional optical system**<br>non-resonant / resonant |
| validity of the electric field | OK / OK | OK / OK |
| validity of the MCE | NG / NG | OK / OK |

### 6.5.3 Under the Resonant Condition

Under the resonant condition, the total electric field is regarded as the cause of response, as is true under the far-field observation condition. Under the resonant condition, $\eta \simeq 1$, the resonant term (the first term) in (6.31) dominates over the non-resonant term (the second term) and asymptotically equals to (6.30).

$$\chi_{ij}^{\mathbf{j}\leftarrow(t)}(\mathbf{r}, \omega) \simeq \chi_{ij}^{\mathbf{j}\leftarrow(\ell)}(\mathbf{r}, \omega)\,. \tag{6.40}$$

This fact together with (6.27) means that the total electric field is considered as the cause of response.

### 6.5.4 Under Non-resonant and NF Incidence and NF Observation Conditions

Otherwise, i.e., under a non-resonant and NF observation condition, the responses to the longitudinal and transverse electric fields are different and dependent on the resonance factor $\eta$ and orbitals. Therefore, the two types of the electric fields should be treated separately as the cause of response using the microscopic single susceptibilities (Table 6.3).

In the context of NFO , the response to the longitudinal electric field is discussed in Chap. 5 in [20] and Chap. 9 in [24]. The present work is a further comparison with the transverse electric field, considering the non-resonant condition. Actually,

**Table 6.4** Comparison of the order carried by the resonant and non-resonant terms in the susceptibilities related to the induced current density, (6.30)–(6.31); the resonant and non-resonant term in (6.31) are labeled by "res", and "non-res", respectively

| | Susceptibility for induced current density |
|---|---|
| $\eta \simeq 0$ ($\hbar\omega \gg \hbar\Delta\omega_1$) | $\left\|\chi_{ij}^{\mathbf{j}\leftarrow(\ell)}(\mathbf{r},\omega)\right\| \simeq \left\|\chi_{ij}^{\mathbf{j}\leftarrow(t)}(\mathbf{r},\omega)_{\text{non-res}}\right\| \gg \left\|\chi_{ij}^{\mathbf{j}\leftarrow(t)}(\mathbf{r},\omega)_{\text{non}}\right\|$ |
| $\eta \simeq 1$ ($\hbar\omega = \hbar\Delta\omega_1$) | $\left\|\chi_{ij}^{\mathbf{j}\leftarrow(t)}(\mathbf{r},\omega)_{\text{res}}\right\| \simeq \left\|\chi_{ij}^{\mathbf{j}\leftarrow(\ell)}(\mathbf{r},\omega)\right\| \gg \left\|\chi_{ij}^{\mathbf{j}\leftarrow(t)}(\mathbf{r},\omega)_{\text{non-res}}\right\|$ |
| $\eta \gg 1$ ($\hbar\omega \ll \hbar\Delta\omega_1$) | $\left\|\chi_{ij}^{\mathbf{j}\leftarrow(t)}(\mathbf{r},\omega)_{\text{res}}\right\| \simeq \left\|\chi_{ij}^{\mathbf{j}\leftarrow(t)}(\mathbf{r},\omega)_{\text{non-res}}\right\| \gg \left\|\chi_{ij}^{\mathbf{j}\leftarrow(\ell)}(\mathbf{r},\omega)\right\|$ |

$\chi_{ij}^{\mathbf{j}\leftarrow(\ell)}$ and the resonant term in $\chi_{ij}^{\mathbf{j}\leftarrow(t)}$ include the resonance factor $\dfrac{1}{\eta^2-1}$, and the non-resonant term in $\chi_{ij}^{\mathbf{j}\leftarrow(t)}$ cariies the factor 1. In the resonant and non-resonant regime, Table 6.4 summarizes the comparison of the contribution from resonant and non-resonant terms in the susceptibilities, (6.30) and (6.31), which is proportional to the corresponding induced current density under the LWA. The susceptibilities related to the induced charge density, (6.27), are equal all over the range of $\eta$, so that, they are omitted from Table 6.4.

The present model is very simple and may be modified in a many-electron system. However, we consider that the difference between the responses to the longitudinal and transverse electric fields should survive even in the many-electron system. Because such difference comes from *the non-relativistic nature* of the system, as is mentioned in Sect. 6.3.2.

From the above considerations, one may infer a guiding principle that the non-resonant condition and simultaneous the NF incidence and NF observation conditions, a non-metallic material brings about NF-specific phenomena that may not be described using MCEs.

### *6.5.5 Comparison Between the Microscopic and Macroscopic Constitutive Equations*

Let us compare the above results with that from the MCE under the LWA:

$$P_i(\mathbf{r},t) = \theta(\mathbf{r}\in\mathcal{V})(\epsilon(\omega)-\epsilon_0)\left(E_i^{(\ell)}(\mathbf{0},t)+E_i^{(t)}(\mathbf{0},t)\right), \tag{6.41}$$

$$\theta(\mathbf{r}\in\mathcal{V}) \equiv \begin{cases} 1, & \mathbf{r} \text{ belongs to the region of the responding material}, \\ 0, & \text{otherwise}, \end{cases} \tag{6.42}$$

where $\mathbf{P}(\mathbf{r},t)$ and $\epsilon(\omega)$ are the polarization and the electric permittivity of the responding material. The step function $\theta(\mathbf{r}\in\mathcal{V})$ guarantees the Maxwell's boundary condition [25, 26].

$$\Delta\rho(\mathbf{r},t) = -\partial_i P_i(\mathbf{r},t)$$
$$= (\epsilon(\omega)-\epsilon_0)\int_{\partial\mathcal{V}} \mathrm{d}^2 s\, \delta^3(\mathbf{s}-\mathbf{r})\, n_i(\mathbf{s})\cdot\left(E_i^{(\ell)}(\mathbf{0},t)+E_i^{(t)}(\mathbf{0},t)\right), \quad (6.43)$$
$$\Delta j_i(\mathbf{r},t) = \partial_t P_i(\mathbf{r},t)$$
$$= \theta(\mathbf{r}\in\mathcal{V})\delta_{ij}(\epsilon(\omega)-\epsilon_0)\left(\dot{E}_j^{(\ell)}(\mathbf{0},t)+\dot{E}_j^{(t)}(\mathbf{0},t)\right), \quad (6.44)$$

where $\partial\mathcal{V}$, $\mathbf{n}$ are the surface of the responding material and the outgoing unit normal vector on the surface. Under the NF observation condition, the MCE does not support the spatial distribution of the sources derived from the microscopic response. Actually, (6.21) and (6.22) considering (6.27), (6.30), (6.31) represents for the rippling spatial distribution in terms of orbitals, while (6.43) and (6.44) for rather simple spatial distribution, i.e., the surface charge and bulk current in terms of delta and step functions, respectively.

On the contrary, under the far-field observation condition, the spatial average of (6.44) corresponds to (6.36) and one may assign

$$\overline{\chi_{ij}^{\mathbf{j}\leftarrow(\ell)}}(\mathbf{r},\omega) = \overline{\chi_{ij}^{\mathbf{j}\leftarrow(t)}}(\mathbf{r},\omega) = \delta_{ij}(\epsilon(\omega)-\epsilon_0), \quad (6.45)$$

$$\text{where}\quad \epsilon(\omega)-\epsilon_0 = \frac{q^2\hbar^2}{m\,\mathcal{V}}\frac{1}{(\hbar\Delta\omega_1)^2-(\hbar\omega)^2} = \frac{q^2}{m\,\omega^2\,\mathcal{V}}\frac{1}{\eta^2-1}. \quad (6.46)$$

The spatial average of (6.43) vanishes using the Gauss's theorem, $\overline{\Delta\rho(\mathbf{r},t)} = -\overline{\nabla\cdot\mathbf{P}(\mathbf{r},t)} = 0$, which corresponds to (6.35). Under the far-field observation condition, the spatial averages of the microscopic susceptibilities and the MCE are consistent under the LWA.

Under the NF observation condition, one should use microscopic susceptibilities, taking into account the dependency on the orbitals of the electron states. As long as the present two-level model, if the two orbitals are known, the microscopic susceptibilities can be reproduced from the macroscopic susceptibilities which is the spatial average of microscopic one. Actually, in (6.27), (6.30) and (6.31), all the terms except the non-resonant term in (6.31) are the product of the factor of the macroscopic susceptibilities (electric permittivity), (6.46), and the factor of the orbitals, and the non-resonant term is simply determined by the ground state orbital. Therefore, the next relationships hold:

$$\chi_j^{\rho\leftarrow(\ell)}(\mathbf{r},\omega) = \chi_j^{\rho\leftarrow(t)}(\mathbf{r},\omega) = (\epsilon(\omega)-\epsilon_0)\frac{1}{\eta}\mathcal{C}_j\frac{\varphi_0(\mathbf{r})\varphi_1(\mathbf{r})}{1/\mathcal{V}} \tag{6.47}$$

$$\chi_{ij}^{\mathbf{j}\leftarrow(\ell)}(\mathbf{r},\omega) = (\epsilon(\omega)-\epsilon_0)\delta_{ij}\frac{\partial_i\varphi_1(\mathbf{r})\varphi_0(\mathbf{r})-\varphi_1(\mathbf{r})\partial_i\varphi_0(\mathbf{r})}{\mathcal{C}_i/\mathcal{V}} \tag{6.48}$$

$$\chi_{ij}^{\mathbf{j}\leftarrow(t)}(\mathbf{r},\omega) = (\epsilon(\omega)-\epsilon_0)\delta_{ij} \left(\eta^2\frac{\partial_i\varphi_1(\mathbf{r})\varphi_0(\mathbf{r})-\varphi_1(\mathbf{r})\partial_i\varphi_0(\mathbf{r})}{\mathcal{C}_i/\mathcal{V}} - (\eta^2-1)\frac{\varphi_0(\mathbf{r})\varphi_0(\mathbf{r})}{1/\mathcal{V}}\right) \tag{6.49}$$

The similar procedure will be available in NFO, if the response is regulated by the excitation between two levels.

### *6.5.6 A Remark on FDTD Calculations*

The MCE in terms of the permittivity has been widely employed to calculate the optical near field using the finite differential time domain (FDTD) method [27]. One may notice that the permittivity in the FDTD method carries a simple spatial dependence and leads to some quantitative error. Actually, the microscopic susceptibilities, for example, (6.27), (6.30), and (6.31), or (6.47)–(6.49) have rippling spatial distributions originating from the orbitals.

The situation is more serious In the case of the NF optical system (I) in Table 6.2 with a non-metallic material under the non-resonant condition, because the concept electric field is not available and it is a logical fallacy to use the MCE. Thus, a novel simulation method is necessary based on the single susceptibility, for example, employing (6.47)–(6.49) for such an electron system that the present two-level model is feasible.

### *6.5.7 Numerical Demonstration of Distinct Responses*

Here the microscopic susceptibilities are visualized by numerical calculation in a simplified system. Under the LWA, the susceptibilities are proportional to the induced charge and current densities. Let us introduce a small-scale cylinder, in which the one-electron two-level system is bounded. The whole size of the cylinder is much smaller than the light wavelength and the radius is much smaller than the height, so that the radius and angular orbitals, $R(r)$ and $\Theta(\theta)$ may not be excited and only the excitation brings about between the two height orbitals, $Z_0(z)$ and $Z_1(z)$, where $r, \theta, z$ are the cylindrical coordinates. The two orbitals are $\varphi_m(\mathbf{r}) = R(r)\Theta(\theta)Z_m(z)$ $(m = 0, 1)$, which are bound states represented by real functions with well-defined spatial parities, even and odd, respectively, as is assumed in this section. Various quantities characterizing this system are summarized in Table 6.5.

**Table 6.5** Two-level orbitals and associated quantities in the system of the numerical calculation. Equation (6.37) is checked of the completeness of the two orbitals; the unity is ideal

| | |
|---|---|
| $Z_0(z)$ | $\sqrt{\frac{2}{a}}\cos\left(\frac{\pi}{a}z\right)$ |
| $Z_1(z)$ | $\sqrt{\frac{2}{a}}\sin\left(\frac{2\pi}{a}z\right)$ |
| $\mathcal{D}_z$ | $\frac{16}{9\pi^2}a$ |
| $\mathcal{C}_z$ | $\frac{16}{3}\frac{1}{a}$ |
| $\frac{Z_1'(z)Z_0(z)-Z_1(z)Z_0'(z)}{\mathcal{C}_z/a}$ | $\frac{3\pi}{4}\cos^3\left(\frac{\pi}{a}z\right)$ |
| $\frac{Z_0(z)Z_0(z)}{1/a}$ | $2\cos^2\left(\frac{\pi}{a}z\right)$ |
| $\mathcal{D}_z\mathcal{C}_z$ (l.h.s of (6.37)) | $\frac{256}{27\pi^2}=0.9607$ |

After the integration of (6.27), (6.30), (6.31) over the radius and angular coordinates, the following susceptibilities that relates the longitudinal and transverse electric fields to the linear charge density and current:

$$
\begin{aligned}
X_z^{\rho\leftarrow(\ell)\mathrm{or}(t)}(z,\omega) &= \frac{q^2}{m\omega^2 a}\frac{\mathcal{C}_z}{\eta(\eta^2-1)}\frac{Z_0(z)\,Z_1(z)}{1/a} \\
&= \frac{q^2}{m\omega^2 a}\frac{\mathcal{C}_z}{\eta(\eta^2-1)}\frac{1}{a}\frac{32}{3}\cos\left(\frac{\pi}{a}z\right)\sin\left(\frac{2\pi}{a}z\right)
\end{aligned}
\tag{6.50}
$$

$$
\begin{aligned}
X_{zz}^{\mathbf{j}\leftarrow(\ell)}(z,\omega) &= \frac{q^2}{m\omega^2 a}\frac{1}{\eta^2-1}\frac{\partial_z Z_1(z)\,Z_0(z)-Z_1(z)\partial_z Z_0(z)}{\mathcal{C}_z/a} \\
&= \frac{q^2}{m\omega^2 a}\frac{3\pi}{4}\cos^3\left(\frac{\pi}{a}z\right)
\end{aligned}
\tag{6.51}
$$

$$
\begin{aligned}
X_{zz}^{\mathbf{j}\leftarrow(t)}(z,\omega) = \frac{q^2}{m\omega^2 a}\Bigg(&\frac{\eta^2}{\eta^2-1}\frac{\partial_z Z_1(z)\,Z_0(z)-Z_1(z)\partial_z Z_0(z)}{\mathcal{C}_z/a} \\
&-\frac{Z_0(z)Z_0(z)}{1/a}\Bigg)
\end{aligned}
\tag{6.52}
$$

where

$$\mathcal{C}_z \equiv \int_{-a/2}^{a/2} \mathrm{d}z\,(\partial_z Z_1(z)\, Z_0(z) - Z_1(z)\partial_z Z_0(z))\,, \tag{6.53}$$

$$\mathcal{D}_z \equiv \int_{-a/2}^{a/2} \mathrm{d}z Z_1(z)\, z\, Z_0(z)\,. \tag{6.54}$$

Under the non-resonant condition in this cylinder model, there appears the difference between the currents induced by longitudinal and transverse electric fields, although it is not drastic, perhaps, due to the simpleness of the cylinder model (Fig. 6.2).

## 6.6 Linear Response Theory of Many-electron System in Near-field Optics

This section develops the linear single susceptibility equally associated with the scalar and vector potentials, satisfying the requisites mentioned in Sect. 6.4.5. In near future, we will give general formulation on the basis of action integral and apply to the nonlinear single susceptibility, which is essential to describe the DP (Fig. 6.3).

### *6.6.1 Notations*

Here is the summary of notations. In the formulation of the semiclassical response theory for the many-electron system in NFO, we utilize the classical scalar and vector potentials, $\phi(x)$ and $\mathbf{A}(x)$, respectively, (as the cause of response), and the electron charge and current densities, $\hat{\rho}(x)$ and $\hat{j}_i(x)$, respectively, in the form of Heisenberg operators (as the result of response), which are defined as:

$$\hat{\rho}(x) \equiv q\hat{\psi}^\dagger_\alpha(x)\hat{\psi}_\alpha(x), \text{ and} \tag{6.55}$$

$$\hat{j}_i(x) \equiv \frac{q}{2m}\hat{\psi}^\dagger_\alpha(x)\left(\frac{\hbar}{i}\partial_i - qA_i(x)\right)\hat{\psi}_\alpha(x) + \text{h.c.}\,, \tag{6.56}$$

where $x = (\mathbf{r}, t)$ represents for the temporal and spatial coordinates, $m$ and $q(= -e)$ are the electron mass and charge, respectively. $\hat{\psi}^\dagger_\alpha$, $\hat{\psi}_\alpha$ are the electron field operators with the spin state $\alpha$ (one of the two spin states; so called "up" and "down" states). In the right-hand side of (6.55)–(6.56), the Einstein's rule is applied, that is, summation should be executed over repeated indices, $\alpha$. The electron field operators are the quantized Grassmann fields satisfying the following anti-commutation relationship:

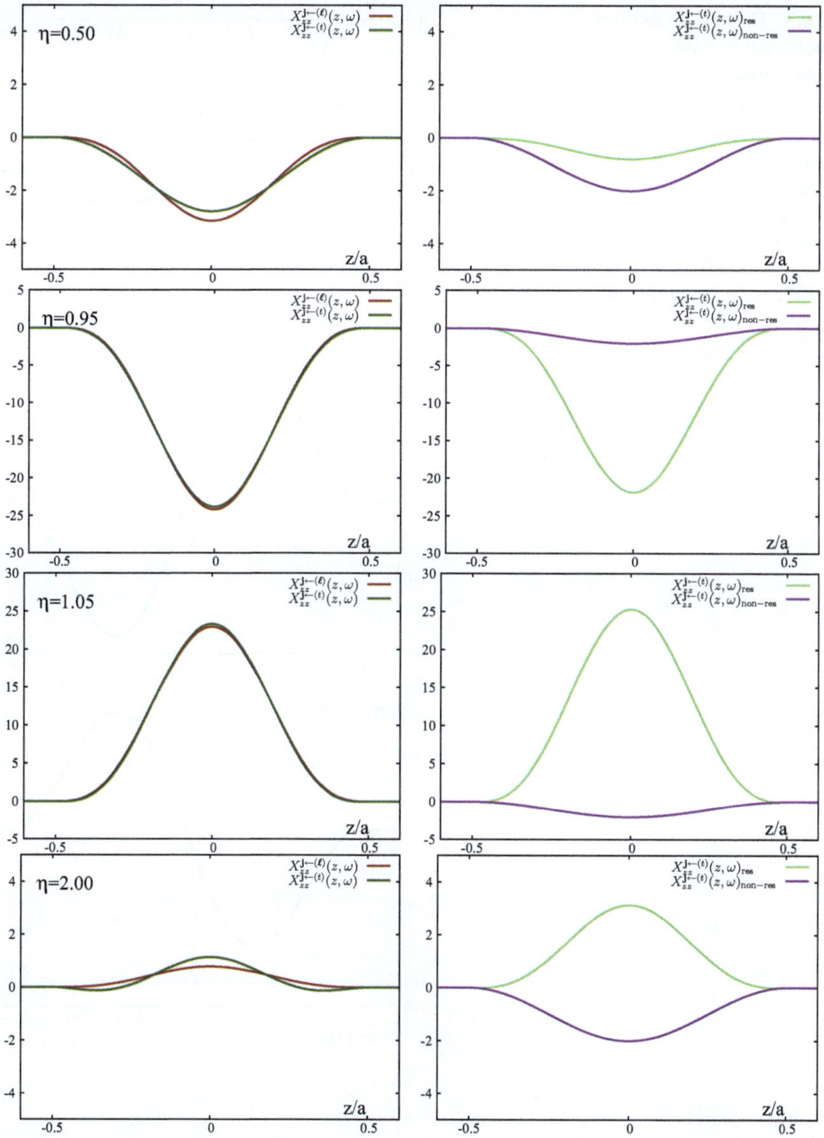


**Fig. 6.2** [Left column] Comparison between the susceptibilities proportional to the induced current density, $\chi_{zz}^{\mathbf{j}\leftarrow(\ell)}(z,\omega)$ (red line) and $\chi_{zz}^{\mathbf{j}\leftarrow(t)}(z,\omega)$ (green line); over the various resonance parameter, $\eta$. The horizontal axis is position; the material lie in the region $z/a \in [-0.5, +0.5]$, and the vertical axis is amplitude in the unit of $\frac{q^2}{m\omega^2 a}$. Under the non-resonant conditions, $\eta = 0.5, 1.5$, the difference of the two become clearer than the near-resonant cases, $\eta = 0.95, 1.05$, although the amplitude is small. [Right column] The contributions to $\chi_{zz}^{\mathbf{j}\leftarrow(t)}(z,\omega)$ from the resonant and non-resonant term (green and purple lines, respectively); the latter is independent with $\eta$, and gives the same profile all over the $\eta$s. The behaviors in these figures are consistent with the Table 6.4

**Fig. 6.3** The susceptibilities proportional to induced charge density, $\chi_z^{\rho\leftarrow(\ell)}(z,\omega) = \chi_z^{\rho\leftarrow(t)}(z,\omega)$, over the various resonance parameter, $\eta$. The horizontal axis is position; the material lies in the region $z/a \in [-0.5, +0.5]$, and the vertical axis is amplitude in the unit of $\frac{q^2}{m\omega^2 a^2}$

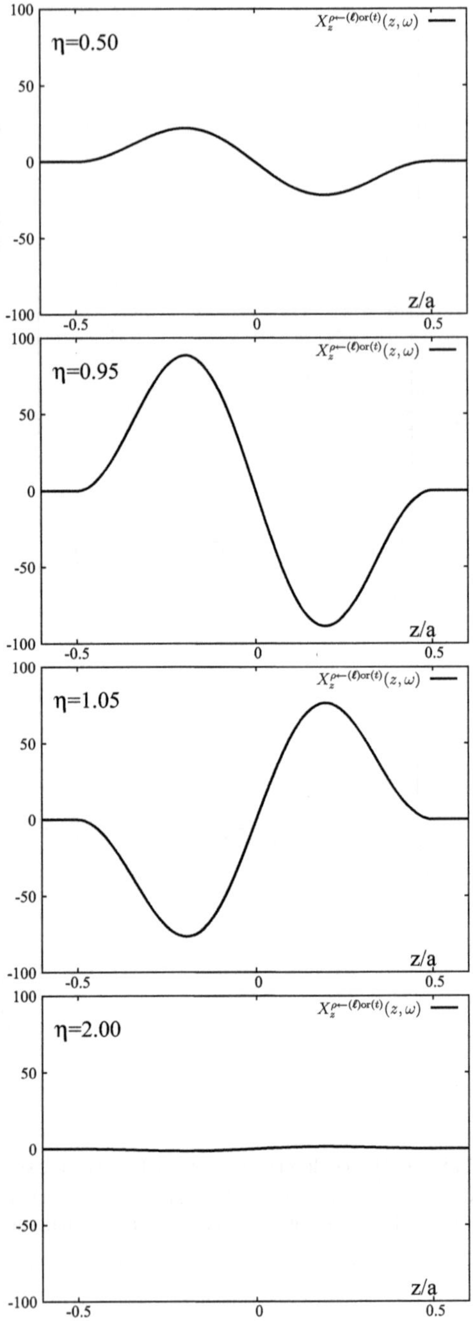

$$[\hat{\psi}_\alpha(\mathbf{r}, t), \hat{\psi}^\dagger_\beta(\mathbf{r}', t)]_+ = \delta^3(\mathbf{r} - \mathbf{r}')\delta_{\alpha\beta} \,. \tag{6.57}$$

In the the 4-element notations, the scalar and vector potentials and the induced charge and current densities are expressed as,

$$A^\mu = (\phi, c\mathbf{A}), \quad A_\mu = (\phi, -c\mathbf{A}), \tag{6.58}$$

$$\hat{j}^\mu = (c\hat{\rho}, \hat{\mathbf{j}}), \quad \hat{j}_\mu = (c\hat{\rho}, -\hat{\mathbf{j}}) \,, \tag{6.59}$$

where $c$ is the speed of light. The derivative operators in the 4-element notation are:

$$\partial^\mu = (1/c\, \partial_t, -\nabla), \quad \partial_\mu = (1/c\, \partial_t, \nabla), \tag{6.60}$$

$$\Box = \partial^\mu \partial_\mu = 1/c^2\, \partial_t^2 - \Delta \,. \tag{6.61}$$

The Einstein rule is applied to the indices of vector as is done for the indices of Grassmann fields.

Although Lorentz invariance is not maintained in the non-relativistic theory of our interest, we use the 4-element notation to simply represent charge conservation and gauge invariance. Actually, the charge conservation law, $\partial_t \hat{\rho}(x) + \partial_i \hat{j}_i(x) = 0$ (one may checked through explicit calculation) is simply expressed in the 4-element notation as,

$$\partial_\mu \hat{j}^\mu(x) = 0. \tag{6.62}$$

Such the quantities as the total Hamiltonian and the total action integral are invariant under the gauge transformation:

$$A^\mu(x) \to A^\mu(x) + c\, \partial^\mu \eta(x),$$
$$\hat{\psi}_\alpha(x) \to e^{\frac{i}{\hbar} q\eta(x)} \hat{\psi}_\alpha(x), \quad \hat{\psi}^\dagger_\alpha(x) \to \hat{\psi}^\dagger_\alpha(x) e^{\frac{-i}{\hbar} q\eta(x)} \,, \tag{6.63}$$

where $\eta(x)$ is the gauge function. From the point of view of Noether's theorem[29, 30], the gauge invariance is the cause of the charge conservation law, (6.62).

### 6.6.2 Definition of Linear Single Susceptibility in the Form of Heisenberg Operator

To formulate the response function in the perturbation theory, let us separate the EM field into two parts:

$$A^\mu(x) = A^{(0)\mu}(x) + \Delta A^\mu(x), \tag{6.64}$$

where $A^{(0)\mu}$ is the EM potential in the non-perturbed system, and $\Delta A^{\mu}(x)$ is the perturbative EM potential, which is considered as the cause of response. The result of response is the induced charge and current densities, which should be expressed by the Taylor expansion with respect to the perturbation EM potential, $\Delta A^{\mu}(x)$. The lowest order expression is,

$$\hat{j}^{\mu}(x;[A^{(0)\nu}+\Delta A^{\nu}]) - \hat{j}^{\mu}(x;[A^{(0)\nu}]) = \int \mathrm{d}^4 x_1 \left. \frac{\delta \hat{j}^{\mu}(x;[A^{\nu}]))}{\delta A^{\mu_1}(x_1)} \right|_{A^{\nu}=A^{(0)\nu}} \Delta A^{\mu_1}(x_1)\,. \tag{6.65}$$

The coefficient of the Taylor expansion is the linear single susceptibility in the Heisenberg-operator form.

$$\hat{\chi}^{\mu}{}_{\mu_1}(x,x_1) \equiv \left. \frac{\delta \hat{j}^{\mu}(x;[A^{\nu}])}{\delta A^{\mu_1}(x_1)} \right|_{A^{\nu}=A^{(0)\nu}} . \tag{6.66}$$

### 6.6.3 *Evaluation of Linear Single Susceptibility of Many-electron System in NFO*

To evaluate the linear single susceptibility defined in (6.66) in the semiclassical manner, we employ the Hamiltonian without eliminating the scalar potential, which should be equally treated as the vector potential, and is adequate for many-electron system in NFO as mentioned in Sect. 6.4.5. The Hamiltonian, $\hat{H}$ is divided into the non-perturbation and perturbation parts, $\hat{H}^{(0)}$ and $\hat{V}(t)$, respectively:

$$\begin{aligned}\hat{H}^{(0)} \equiv \int \mathrm{d}^3 x\, \frac{1}{2m}\left(\frac{\hbar}{-i}\partial_i - qA_i^{(0)}(x)\right)\hat{\psi}_{\alpha}^{\dagger}(x)\cdot\left(\frac{\hbar}{i}\partial_i - qA_i^{(0)}(x)\right)\hat{\psi}_{\alpha}(x) \\ + q\phi^{(0)}(x)\,\hat{\psi}_{\alpha}^{\dagger}(x)\hat{\psi}_{\alpha}(x) + v^{(\mathrm{AUX})}(x)\,\hat{\psi}_{\alpha}^{\dagger}(x)\hat{\psi}_{\alpha}(x),\end{aligned} \tag{6.67}$$

$$\begin{aligned}\hat{V}(t) &\equiv \hat{H} - \hat{H}^{(0)} \\ &= \int \mathrm{d}^3 x \left\{ \frac{1}{c}\Delta A^{\mu}(x)\hat{j}_{\mu}(x) - \frac{q}{2mc^3}\tilde{\delta}_{\mu}^{\mu'}\Delta A^{\mu}(x)\Delta A_{\mu'}(x)\hat{j}_0(x) \right\},\end{aligned} \tag{6.68}$$

$$\text{where } \tilde{\delta}_{\mu}^{\mu'} = \begin{cases} 1 & \text{for } \mu=\mu'=1,2,3 \\ 0 & \text{otherwise} \end{cases} . \tag{6.69}$$

The static auxiliary potential, $v^{(\mathrm{AUX})}(x)$ effectively represents for the quantum many-electron effect (the exchange-correlation effect) ; this fact will be explained in the next subsection, Sect. 6.6.5. At this first stage of investigation, the interaction between spin polarization and the EM field is ignored.

Due to *the non-relativistic nature* of the system, the Hamiltonian depends on the scalar and vector potentials in a different manner. Therefore, the perturbation Hamiltonian includes not only the first order term of $\Delta A^{\mu}(x)$ but also additional second order term, to which only the vector-potential components may contribute; this non-relativistic effect is represented by the tensor (6.69) . In the linear response treatment in Sect. 6.5, this non-relativistic nature has caused the different responses to the longitudinal and transverse electric field leading to the irrationality of the concept "electric field".

On may fomulate the the linear single susceptibility in ordinary linear response theory. As an extensible formulation to the nonlinear single susceptibility, the procedure to derive the linear single susceptibility is performed in two steps. As the first step, the 4-element induced current density is expanded by the perturbation series of $\hat{V}(t)$ within the Heisenberg picture, then, in the next step, the linear susceptibility is derived by the functional derivative of 4-element induced current density with respect to the EM potential. The detail will submitted in the future paper [35], and here the result is:

$$\hat{\chi}^{\mu}{}_{\mu_1}(x, x_1) = \left.\frac{\delta \hat{j}^{\mu}(x)}{\delta A^{\mu_1}(x_1)}\right|_{A=A^{(0)}}$$
$$= \frac{-q}{mc^2}\,\tilde{\delta}^{\mu}{}_{\mu_1}\,\delta^4(x-x_1)\,\hat{j}^{(in0)\,0}(x) + \frac{1}{i\hbar c^2}\theta(ct-ct_1)\left[\hat{j}^{(in0)\,\mu}(x),\,\hat{j}^{(in0)}{}_{\mu_1}(x_1)\right], \tag{6.70}$$

where $\hat{j}^{(in0)\,\mu}(x) = \left.\hat{j}^{(in)\,\mu}(x)\right|_{A=A^{(0)}}$.

### 6.6.4 Charge Conservation Law and Gauge Invariance

As pointed out in Sect. 6.4.5, the linear single susceptibility, (6.70) itself should be responsible for the charge conservation law, (6.62) and gauge invariance under the transformation in (6.63). Actually, this facts are respected in (6.70) as follows.

Considering the expression for the 4-element induced current density in (6.65), the charge conservation law, (6.62) is described by the derivative of the coordinate for the result, $x$:

$$\partial_{\mu}\hat{\chi}^{\mu}{}_{\mu_1}(x, x_1) = 0. \tag{6.71}$$

One can easily check (6.71).

On the other hand, gauge invariance is described by the derivative of the coordinate for the cause, $x_1$:

$$\partial^{\mu_1}\hat{\chi}^{\mu}{}_{\mu_1}(x, x_1) = 0. \tag{6.72}$$

To confirm this fact, consider the convolution integral with the single susceptibility and the perturbative EM field, in a certain gauge, e.g.,

$$\int d^4x_1 \ \hat{\chi}^{\mu}{}_{\mu_1}(x, x_1)\Delta A^{\mu_1}(x_1). \tag{6.73}$$

A gauge transformation of $\Delta A$ to $\Delta A'$ in another gauge is expressed as :

$$\Delta A^{\mu_1}(x_1) = \Delta A'^{\mu_1}(x_1) + c\,\partial^{\mu_1}\eta(x_1), \tag{6.74}$$

where $\eta$ is the gauge function. Equation (6.73) leads to:

$$\begin{aligned}
&\int d^4x_1 \ \hat{\chi}^{\mu}{}_{\mu_1}(x, x_1)\Delta A^{\mu_1}(x_1) \\
&= \int d^4x_1 \ \hat{\chi}^{\mu}{}_{\mu_1}(x, x_1)\Delta A'^{\mu_1}(x_1) - c\int d^4x_1 \ \partial^{\mu_1}\hat{\chi}^{\mu}{}_{\mu_1}(x, x_1)\eta(x_1) \\
&= \int d^4x_1 \ \hat{\chi}^{\mu}{}_{\mu_1}(x, x_1)\Delta A'^{\mu_1}(x_1).
\end{aligned} \tag{6.75}$$

The contribution of the gauge function, $\eta(x)$ vanishes in the convolution integral. Thus, the gauge of the perturbative EM field may be freely selected. This means that the induced charge and current densities are determined independent of the chosen gauge and, in practice, one may select a gauge that is most convenient for calculation. Therefore, (6.72) properly represents for the gauge invariance.

Actually, one can easily check (6.72). Furthermore, if one notices that the symmetry of the coordinates $x$ and $x_1$ in (6.70), one may agree with the reciprocity relationship between (6.71) and (6.72).

### *6.6.5 Preparation for the Ground State using Density Functional Theory*

The actual linear single susceptibility is the expectation values of the corresponding operator, (6.70) using the ground state in the non-perturbed electron system, which is specified by the simplified conditions in this chapter:

$$\mathbf{A}(x) = \mathbf{A}^{(0)}(x) = \mathbf{0}, \tag{6.76}$$

$$\phi(x) = \phi^{(0)}(x) \quad \text{is static.} \tag{6.77}$$

The density functional theory (DFT) [31] may allow us to prepare such the ground state and the complete set of the states in the electron system. The fundamental idea of the Kohn–Sham scheme [32] in the DFT is that the auxiliary potential, $v^{(\mathrm{AUX})}(\mathbf{r})$ adjusts the non-interacting electron charge density, to the proper interacting electron charge density, which is evaluated under the ordinary Hamiltonian with eliminating the scalar potential. Actually, the present non-perturbed Hamiltonian, (6.67) in the semiclassical treatment does not take into account for the quantum many-electron effects (the exchange and correlation effects), so that (6.67) is considered as that for non-interacting electron gas in the Kohn–Sham scheme in the DFT, if the auxiliary potential, $v^{(\mathrm{AUX})}(\mathbf{r})$ is assigned as the exchange-correlation potential.

Therefore, the ground state $|0\rangle$ in need may be determined by the Kohn–Sham scheme in the DFT. For details, one may prepare the spin-orbit function $\varphi_{k\alpha}(\mathbf{r},\sigma)$ ($k,\alpha$ stands for the orbit and spin states, $\sigma$ is the spin coordinate) as the eigenstate of the Kohn–Sham equation [Equation (2.8) in [32]] with the eigenenergy $\hbar\omega_k$. Under the conditions,(6.76)–(6.77), the Kohn–Sham equation is,

$$0=\left(\hbar\omega_k-q\phi^{(0)}(\mathbf{r})-\frac{1}{2m}\frac{\hbar}{i}\partial_i\cdot\frac{\hbar}{i}\partial_i-v^{(\mathrm{AUX})}(\mathbf{r})\right)\varphi_{k\alpha}(\mathbf{r},\sigma), \tag{6.78}$$

where $v^{(\mathrm{AUX})}(\mathbf{r})$ is set to the exchange-correlation potential. Using these spin-orbit, the electron field operator is expanded as,

$$\hat{\psi}_\alpha(x)=\sum_k\varphi_{k\alpha}(\mathbf{r},\sigma)\hat{a}_{k\alpha}\,, \tag{6.79}$$

where $\hat{a}_{k\alpha}$ is the operator to annihilate the electron belonging to the spin-orbit $\varphi_{k\alpha}(\mathbf{r},\sigma)$. Then, the ground state with the electron number $n$ in need of our theory is expressed as the single Slater determinant,

$$|0\rangle=\frac{1}{\sqrt{n!}}\prod_{k\alpha}\hat{a}^\dagger_{k\alpha}|\mathrm{vac}\rangle\,, \tag{6.80}$$

where $|\mathrm{vac}\rangle$ is the vacuum state, and the indecies $k\alpha$ scan over the $n$ spin-orbits from the lowest eigenenergies. Furthermore, under the fixed $v^{(\mathrm{AUX})}(\mathbf{r})$ and $\phi^{(0)}(\mathbf{r})$, one may consider all the possible combination of $n$ spin-orbits and obtain the normalized orthogonal complete set $\{|m\rangle|m=0,1,2,\ldots\}$ in terms of single Slater determinants.

The expectation value of the linear single susceptibility operator is,

$$\begin{aligned}\langle 0|\hat{\chi}^{\mu}{}_{\mu_1}(x,x_1)|0\rangle =&\frac{-q}{mc^2}\,\tilde{\delta}^{\mu}{}_{\mu_1}\delta^4(x-x_1)\,\langle 0|\hat{j}^{(in0)\,0}(x)|0\rangle \\ &+\frac{1}{i\hbar c^2}\theta(ct-ct_1)\langle 0|\left[\hat{j}^{(in0)\,\mu}(x),\hat{j}^{(in0)}_{\mu_1}(x_1)\right]|0\rangle\,.\end{aligned} \tag{6.81}$$

Next, to evaluate the products of two (or more) current density operators, e.g., the second term in (6.81), we may use the projection operator $\hat{1}=\sum_m|m\rangle\langle m|$. If the convolution integral with the perturbative EM field is performed, the energy denominator will appear. It is important that the irrationality of total electric field that appeared in one-electron NF optical system (I), is also true in the many-electron system because the linear susceptibilities in many-electron system, (6.81), and those in one-electron system included in (6.17)–(6.18) are in the same form, owing to the DFT. As a result, the irrationality of the total electric field originates form *the non-relativistic nature* of the system and survives in the many-electron system in NFO.

### 6.6.6 Summary of Linear Single Susceptibility of Many-electron System

The present linear single susceptibility relates the EM potential (as the cause) to the induced charge and current densities (as the result), and is derived in the form of Heisenberg operator by the functional derivative of the induced current density with respect to the EM potential, and, furthermore, guarantees the charge conservation law and the gauge invariance.

It is suggested that the ground state is determined by means of the DFT to evaluate the expectation value of the linear single susceptibility operator. Owing to the DFT, the scalar potential inherently existing in the electron system is separated as $\phi^{(0)}(x)$ and $v^{(AUX)}(\mathbf{r})$, and the scalar potential incident may be treated equally with the vector potential incident. This fact satisfies one of the requisites mentioned in Sect.6.4.5. Note that, $\phi^{(0)}(x)$ is under the Coulomb gauge but the scalar and vector potential incidents may be gauge-free, that is, the present response theory is still free from gauge-fixing.

The logical fallacies lie in the familiar concepts such as the electric field, permittivity, and polarization remain in the many-electron system in the same manner as in the one-electron system.

## 6.7 Conclusions

1. The redefinition of the DP is given based on the Heisenberg's uncertainty principle , without knowing the details of the interaction associated with the responding material.
2. The NF optical system is characterized and classified by the positions of the light source and the light sink . Furthermore, the system the DP appears is assigned as the limit of the NF optical system, in which the light source and light sink is close to the localized EM field of interest.
3. The strategy of approach by means of response theory is explained using the Thomas–Fermi theory as a guide.
4. There are two main difficulties to develop the theory for optical near field and the DP. Firstly, it is logical fallacy in NFO to use the familiar concept such as the electric field, electric permittivity, and polarization. Therefore, it is essential

to use the EM potential instead of the electric and magnetic fields, the induced charge and current densities instead of the polarization and magnetization, and the single susceptibility instead of the electric permittivity and magnetic permeability. This logical fallacy originates in the non-relativistic nature of the system, and comes into observation only in the restricted NF optical system, that is, the system with the non-metallic material under non-resonant condition and the NF incidence and NF observation conditions; this is demonstrated in one-electron system. On the contrary, in conventional optical systems, such are under the far-field observation condition and/or the resonant condition, the familiar concepts function as expected.

5. The other difficulty is that the NFO is inevitably related to the many-electron problem. It is suggested to use the Hamiltonian without eliminating the scalar potential for the equal treatment of the scalar and vector potentials as the cause of response, and derive linear single susceptibility. in the form of Heisenberg operator. It is also suggested that the ground state is determined by the Kohn–Sham scheme in the DFT, and may be employed to take the expectation value of the linear single susceptibility operator. In this procedure, the scalar potential inherently existing in the electron system is separated as $\phi^{(0)}(x)$ and $v^{(\mathrm{AUX})}(\mathbf{r})$, and the scalar potential incident may be treated equally with the vector potential incident.
6. The irrationality of the electric field, permittivity, and polarization originate from the non-relativistic nature of the system, and it is described in the same manner in the many-electron system, owing to density functional theory.
7. The linear single susceptibility relates the EM potential and the induced charge and current densities and guarantees the charge conservation law and the gauge invariance.
8. The DP is strongly coupled with the responding material, so that the nonlinear response should be take into account. Further understanding on nonlinear and non-resonant terms in the susceptibilities [35] are essential to grasp the physics of the DP.

**Acknowledgements** The author are grateful Drs. M. Ohtsu [Research Origin of Dressed Photon (RODreP)], I. Ojima (RODreP), H. Sakuma (RODreP), H. Saigo (Nagahama Institute of Bi-Science and Technology), K. Okamura (Nagoya Univ.), H. Ando (Chiba Univ.), and T. Kawazoe (Tokyo Denki Univ.) for useful discussions on the context of dressed photon, and thanks Dr. K. Cho in Osaka Univ. for useful discussions concerning susceptibility. This work is partially supported by JSPS KAKENHI Grant Number JP25610071 during 2013-2015, and Research Foundation for Opto-Science and Technology during 2018-2019.

## References

1. T. Kawazoe, Y. Yamamoto, M. Ohtsu, Appl. Phys. Lett. **79**, 1184 (2001)
2. H. Yonemitsu, T. Kawazoe, K. Kobayashi, M. Ohtsu, J. Photolumin. **122**, 230 (2007)
3. T. Yatsui, K. Hirata, W. Nomura, Y. Tabata, M. Ohtsu, Appl. Phys. B **93**, 55 (2008)

4. T. Kawazoe, H. Fujiwara, K. Kobayashi, M. Ohtsu, Quantum electronics. IEEE J. Sel. Top. **15**, 1380–1386 (2009)
5. H. Fujiwara, T. Kawazoe, M. Ohtsu, Appl. Phys. B **100**, 85 (2010)
6. T. Kawazoe, K. Kobayashi, M. Ohtsu, Appl. Phys. Lett. **86**, 103102–1 (2005)
7. T. Kawazoe, M. Ohtsu, S. Aso, Y. Sawado, Y. Hosoda, K. Yoshizawa, K. Akahane, N. Yamamoto, M. Naruse, Appl. Phys. B **103**, 537 (2011)
8. T. Kawazoe, M.A. Mueed, M. Ohtsu, Appl. Phys. B **104**, 747 (2011)
9. T. Kawazoe, M. Ohtsu, K. Akahane, N. Yamamoto, Appl. Phys. B **107**, 659 (2012)
10. M. Ohtsu, *Silicon Light-Emitting Diodes and Lasers* (Springer International Publishing, Switzerland, 2016)
11. N. Tate, T. Kawazoe, W. Nomura, M. Ohtsu, Sci. Rep. **5**, 12762–1 (2015)
12. N. Tate, T. Kawazoe, S. Nakashima, W. Nomura, M. Ohtsu, in *Abstracts of the 22nd International Display Workshops*, Otsu, Japan, PRJ3-1, 9–11 Dec 2015
13. M. Ohtsu, *Dressed Photons: Concepts of Light-Matter Fusion Technology* (Springer, Berlin, Heidelberg, 2014)
14. T. Kawazoe, K. Kobayashi, S. Takubo, M. Ohtsu, J. Chem. Phys. **122**, 024715–1 (2005)
15. K. Kobayashi, T. Kawazoe, M. Ohtsu, IEEE Trans. Nanotech. **4**, 517 (2005)
16. N. Wada, M.A. Tran, T. Kawazoe, M. Ohtsu, Appl. Phys. A **115**, 113 (2014)
17. M. Yamaguchi, T. Kawazoe, M. Ohtsu, Appl. Phys. A **115**, 119 (2014)
18. T. Kawazoe, K. Nishioka, M. Ohtsu, Appl. Phys. A **121**, 1409 (2015)
19. J.M. Ziman: *Principles of the Theory of Solids*, 2nd edn. (Cambridge University Press, Cambridge, 1972)
20. K. Cho, *Reconstruction of Macroscopic Maxwell Equations* (Springer, Berlin, Heidelberg, 2010)
21. K. Cho, J. Phys. Condens. Matter **20**, 175202 (2008)
22. K. Cho, *Optical Response of Nanostructures* (Springer, Berlin, Heidelberg, 2003)
23. I. Ojima, *Quantum Field and Micro-Macro Duality* (in Japanese) (Maruzen Shuppan, 2010) (Ryoushiba To Mikuro· Makuro Soutsuisei)
24. O. Keller, *Quantum Theory of Near-Field Electrocynamics* (Springer, Berlin, Heidelberg, 2011)
25. I. Banno: Phys. Rev. A **77**, 033818–1 (2008) ; Erratum *ibid.* **84**, 019906(E) (2011)
26. I. Banno, K. Fujima, Phys. Rev. A **78**, 033816–1 (2008)
27. K. Yee, IEEE Trans. Antennas Propag. **14**, 302 (1966)
28. T. Kugo, *Quantum Theory of Gauge Fields* (in Japanese) (Baifukan, Tokyo, 1989), Chapter 2 (Gejiba No Ryoushiron)
29. The Mathematical Society of Japan and K. Ito eds. *Encyclopedic Dictionary of Mathematics* 2nd edn. (MIT Press, Cambridge, 1993), 318.B; This book was translated from Iwanami Sūgaku Jiten 3rd edn. (Iwanami Shoten, 1985)
30. T. Kugo, [30] Chap. 1 (Gejiba No Ryoushiron)
31. P. Hohenberg, W. Kohn, Phys. Rev. **136**, 3864 (1964)
32. W. Kohn, L.J. Sham, Phys. Rev. **140**, A1133 (1965)
33. Y. Toyozawa, in *The Physics of Elementary Excitations*, ed. by S. Nakajima, Y. Toyozawa, R. Abe (Springer, Berlin, Heidelberg, 1980), Chapter 2, This book was translated from Japanese book "Bussei II" in Iwanami Series of Fundamental Physics (1978)
34. K. Cho, *Optical Response of Nanostructures* (Springer, Berlin, Heidelberg, 2003)
35. K. Nishijima, *Field Theory* (in Japanese) inokuniya-Shoten, Chap. 5 (Tokyo, 1987)

# Index




T. Yatsui (ed.), *Progress in Nanophotonics 5*, Nano-Optics and Nanophotonics,
https://doi.org/10.1007/978-3-319-98267-0

Q



R



S



T



U



V



W

MIX
Papier aus verantwortungsvollen Quellen
Paper from responsible sources
FSC® C105338

If you have any concerns about our products,
you can contact us on
**ProductSafety@springernature.com**

In case Publisher is established outside the EU,
the EU authorized representative is:
**Springer Nature Customer Service Center GmbH**
**Europaplatz 3, 69115 Heidelberg, Germany**

Printed by Libri Plureos GmbH
in Hamburg, Germany